The Transistor

A Grain of Sand that Changed the World

GEOFFREY EVANS C.Eng.

Fellow of the Institution of Engineering & Technology

Redwoods Publishing
Digswell
Hertfordshire
England
www.redwoodspublishing.com

First published in Great Britain 2016
Reprinted 2017

ISBN 978-0-9547455-5-4

Cover design by Martin Stephen Arts

Printed by Print Resources
Welwyn Garden City
Hertfordshire
England

Contents

Plates

Acknowledgements

I was urged to write this book for fear that history will remember the transistor as a form of portable radio instead of the world-changing tiny chip of silicon that it really is.

I owe much to Barbara, my wife, for her patience and encouragement during this time-consuming project.

My thanks to Wikipedia and Image websites for historic information and some of the pictures. References to a number other sources for which I am grateful, are mentioned in the footnotes.

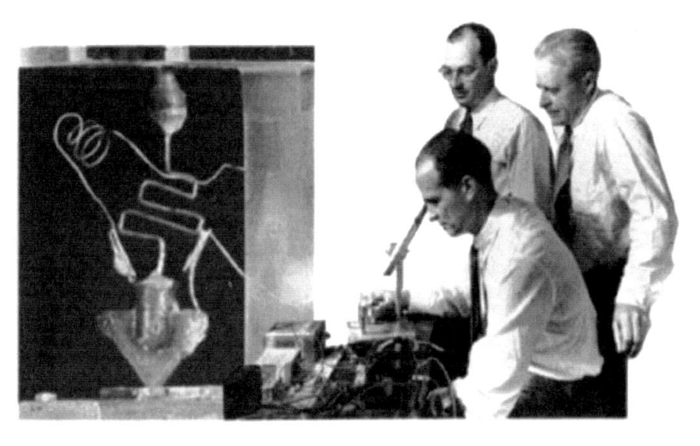

1947: The invention of the Transistor
William Shockley, John Bardeen, Walter Brattain and the first point contact transistor built by Walter Brattain.
(http://www.cedmagic.com/selectavision.html)

Preface

Travelling in our car, guided by our satnav, Barbara answered a call on her mobile phone. It was from our grandson in Zimbabwe. His voice was loud and clear. Arriving home, we switched on our high definition flat screen colour television to hear the news. It showed bright, distinct, moving pictures of a correspondent in a remote part of the world at that very point in time. Then, pleased to be home, we adjusted the automatic controls on our central heating and cooked a meal in our microwave oven before checking our emails on the computer.

At the time, we gave no thought to the fact that none of this would have been possible without the transistor, an incredible device invented as recently as 1947 AD. This minute crystal changed the world forever and yet few of the seven billion members of the population of the world are aware of its existence.

A new millennium popularity poll took place between 2,000 and 2,007 to choose 'The New Seven Wonders of the World'. Selections were from 200 existing monuments. If we were to ask today, "What were the greatest *technological* wonders of the 20th century" the selection is likely to include products like television, MRI scanners, computers, smart phones, the Internet and jumbo jets. However, that would be like choosing the chickens not the eggs. It was the transistor, a minute crystal of a semi-conductor derived from sand, a metalloid called silicon, which made these advances possible.

It is astonishing that, in a period of just over sixty-five years, the radio valve, the size of an electric light bulb, has become replaced with a component so small that six million will fit on a full stop, like the one at the end of this sentence.

The introduction of the internal combustion engine has been described as "transformational giving an unprecedented level of intensity"[1]. Surely, the transistor is at least as transformational and it too has caused an unprecedented level of intensity.

It is unfortunate that manufacturers adopted the word 'transistor' to popularise the portable radio receiver because the latter is really just another small chicken in a vast hencoop. I have tried here to put the transistor in its rightful place and demonstrate that its invention was one of the great turning points in the history of the world; arguably one that is more influential than any other 'wonder'.

Without the transistor, many of our great industries such as Intel, AMD, Apple, ARM, Canon, IBM, LG, Microsoft, Panasonic, Samsung, Sony, and Vodafone, would never have existed

It is hard to find any human activity, whether it is on the earth, in and under water or in space, which does not depend upon, and benefit from, the transistor and its progenies. The list is so long that I have had to confine this work to a small selection; ever conscious that much has been left out. Otherwise it would have become as large as the Encyclopaedia Britannica or Wikipedia; itself another of the transistor's chickens.

In Britain, the engineering profession has suffered from a problem of image and public perception, as illustrated by the following entry in Wikipedia.[2]

> British school children in the 1950s were brought up with stirring tales of "the Victorian Engineers", chief amongst whom were the Brunels, the Stephensons, Telford and their contemporaries. In the UK, "engineering" was more recently perceived as an industry sector consisting of employers and employees loosely termed "engineers" who included the semi-skilled trades. However, the 21st-century view, especially amongst the more educated members of society, is to reserve the term Engineer to describe a

---

1 Jeremy Warner Daily Telegraph 6th January 2016.

2 http://en.wikipedia.org/wiki/Engineer

> university-educated practitioner of ingenuity represented by the Chartered (or Incorporated) Engineer. However, a large proportion of the UK public still sees Engineers as semi skilled tradespeople with a high school education.

I make no apology therefore for my related comments in the text, particularly in the text box on page 60. My hope is that 'The Transistor' will help to put the profession into proper context and encourage young readers to take up electronic engineering as a career.

A father told his teenage child, "If you choose engineering for your living you *may* not become fabulously rich but you will always have a job and never dread Monday mornings." That was certainly my experience, and judging by what we read in Chapter 5, I cannot imagine a more exciting way to earn a living. Then there is always the possibility that, like Bill Gates, you could even become fabulously rich.

The book is set out for two levels of reader. It gives the choice of a quick read or one with added text in boxes for those who may want to dig deeper.

We will encounter some huge, mind numbing numbers as we read on, so it will be useful to have the following definitions to hand.

One kilo = 1,000. (10^3)
One mega = 1,000,000 (10^6)
One giga = 1,000,000,000. (10^9)
One tera = 1,000,000,000,000. (10^{12})
One peta = 1,000,000,000,000,000. (10^{15})
One exa = 1,000,000,000,000,000,000. (10^{18}).
One zetta = 1,000,000,000,000,000,000,000. (10^{21})
One yotta = 1,000,000,000,000,000,000,000,000. (10^{24})
One bronto = 1,000,000,000,000,000,000,000,000.000 (10^{27})
One geop = 1,000,000,000,000,000,000,000,000.000.000 (10^{30})

Geoffrey Evans

Chapter 1. Invention, Discovery, Experiment and Luck

A number of spectacular discoveries and inventions date-mark the history of mankind. We call them 'ages'. They behave like evolutionary milestones of such significance that they change the course of history. Fire, stone, bronze and iron, each earned the accolade of 'an age' and it is possible that the invention of the transistor will give us the *transistor* (or *solid-state*) age.

Inventions and great works of art are rarely spontaneous. They are evolutionary, building on the works of predecessors. They are crowning achievements based on foundations built over many earlier years of trial and error.

For instance, the awesome Airbus 380 aircraft would not have been possible without the work of the Wright Brothers or the evolution of carbon fibre. Galileo Galilei required the prior work of Copernicus and the availability of suitable glass and metals to allow him to build a telescope. Richard Trevithick produced a steam engine in 1804 that ran on rails. Yet, it was George Stephenson who built a viable railway system in 1825 and who we most remember.

Other discoveries are due to a lucky event. Alexander Fleming's luck in 1928 was an unexpected mould growth on a petri dish of jelly. His genius was to question his observation and evolve an antibiotic that he called Penicillin.[3] Chad Orzel, in his book 'How to Teach Quantum Physics to Your Dog'[4], recalls that Davisson and Germer got the idea that a material particle, such as an electron, ought to have a wavelength only because they broke a piece of their apparatus.

---

3 Prompted by the name of the mould, Penicillium rubens.

4 One World Publications 2010. ISBN 978-1-85168-779-4. Page 34.

Our story starts with James Clerk Maxwell and his work on electromagnetism. Many others scratched the surface of wireless telegraphy and the use of radio waves, but Maxwell's genius gave it scientific standing. Nevertheless, progress from his laboratory demonstrations to useful commercial products such as radio and television, spanned several decades and required the efforts of many other workers in the field. History is strewn with examples of inventions that have remained in the shade until someone promoted them commercially.

Many famous men and women of art and science have depended on the efforts of lesser-known workers (often working alone with few resources) who themselves, depended on those that came earlier to pave *their* way. However, this does not detract from the qualities of any of those in the chain of events. It is just that, as time progressed, knowledge accumulated and more sophisticated tools and materials became available for them to use.

Less attractive is the fact that history is beset with inventors who remain unknown merely because others copied their discoveries and publicised them first.

Because knowledge built on knowledge, progress gathered pace at an unprecedented rate from the middle of the nineteenth century. The lives of the peoples of the world have been changed by discoveries that their forebears would never have imagined. There were so many of them that future historians will have difficulty in deciding whether 'steam', 'electricity', IT, 'communications', 'transportation' and 'space' are contenders for the accolades of 'age'.

By the middle of the twentieth century, progress in electronics was held back by the need for a new and revolutionary device. It was like a pressure cooker waiting to be released. The answer came when earlier work with semiconductor materials led to the invention of the transistor in 1947. It laid the foundation for the computer and other incredibly complex and extremely reliable solid-state devices and gave impetus to further understanding of the mysteries of Quantum Physics.

Surely then, the prime contender for the accolade of 'age' must be *'transistor'* or *'solid state electronics'*?

To start our journey, we need to look into the world of *waves*. Those who are well versed in the subject may wish to skip the next section but it is such an important subject that I hope other readers will not be deterred.

Chapter 2. Life Before the Transistor

Waves

"Waves is Energy", as my science teacher told us at every opportunity; much to the displeasure of the English language department.

The 2004 tsunami (harbour wave) travelled several hundred miles at up to 450 miles per hour from the epicentre of an undersea earthquake. It then hit the Indonesian coast killing 227,000 people. Yet almost no water travelled with it until it reached the shallow waters of the coast. That is the nature of waves; they transfer energy, not matter.

Children demonstrate the nature of sound waves every time they use two tin cans linked with a taut string. The waves of their voices travel from one can to the other yet the string remains in place; no amount of matter has moved along it.

The waves of a tsunami and of the transmission of sound are called 'elastic waves'. Such waves need a medium like a solid material, water or air in order to travel. One may remember the demonstration at school with a bell ringing in a glass jar. When the air was sucked out of the jar the bell became silent. This proved that the sound (elastic) waves needed air molecules to act as a medium.

Electromagnetic waves, such as those used for radio transmissions and for the radiation of heat, light and X-Rays are different. Unlike elastic waves they travel most efficiently through a vacuum such as space. Air, water and buildings tend to absorb them.

Great minds still have questions to answer about the phenomenon of electromagnetism. However, without it there would be no radio, no television and no *us*. We must therefore spend a few moments to look at what we do know about the subject.

Clerk Maxwell

Scottish mathematical physicist, James Clerk Maxwell was a born in 1831. At the age of sixteen he commenced three years at Edinburgh University to study physics, mathematics, and philosophy. It is said that he found the courses rather easy.

Maxwell was only 34 when he demonstrated that electromagnetic (electric and magnetic) fields travel through space as waves, moving at the speed of light. He then predicted the propagation of electromagnetic radio waves in 1873. Then, just four years later, Heinrich Hertz made the first demonstration of the transmission of radio waves through space. Progress was moving fast.

Like sound waves, electromagnetic waves also transfer energy but not matter. This is just as well otherwise, as well as light waves arriving from outer space, lumps of stars would also bombard us.

Electromagnetic waves lose their intensity according to the 'Inverse Square Law'. This means that, if the distance doubles, the intensity reduces by a quarter. (Or, if the distance quadruples, the intensity reduces by a sixteenth and so on). It is astonishing therefore that galaxies, trillions of miles away, are so bright that their light is observable to us here on earth.

The world of electromagnetic waves involves unimaginably large and small numbers. For instance, at the extreme end of the electromagnetic spectrum, they can change direction (from crest to trough) as fast as 300,000,000,000,000,000,000,000,000[5] times (cycles) per second.

A radio signal changes direction (oscillates) at up to 300,000,000 times each second. Yet this is insignificant when compared to visible light which does so at 10,000,000,000,000,000 times each

---

5 300 x 10^{24} hertz or 300 yottahertz

second. It gives the mind the same problem as trying to envisage the massive pistons of a motor car engine going up and down at over 40 times each second as we drive along the motorway and doing so over 200 million times during the lifetime of a modern car.

Heinrich Hertz

We use the word hertz for the International unit of frequency, meaning the number of reversals in amplitude (or 'cycles') per second. The word is named after Heinrich Rudolf Hertz the great German physicist. Heinrich was born in February 1857 in Hamburg, Germany. He was the son of wealthy parents. His father, Gustav Ferdinand Hertz, was a lawyer and later a senator. Heinrich studied sciences and engineering and gained a PhD from the University of Berlin in 1880.

Just four years after James Maxwell gave his demonstration of the transmission of radio waves through space, Heinrich Hertz proved that electricity could be transmitted as electromagnetic waves. By building an apparatus that produced and detected the VHF/UHF[6] radio waves, he became the first person to successfully demonstrate the presence of electromagnetic waves. Hertz's work, therefore, made possible the development of radio, television and radar.

Each of the strings of a piano oscillate (vibrate) at different *sound* frequencies. They range from 27.5 times per second for the long and thick strings at the extreme left of an 88-note keyboard to 4186.01 times per second for the, thin ones at the extreme right. Similarly, electromagnetic waves oscillate at different *electromagnetic* frequencies but in this case, they range from 1,000 times per second to 300 followed by 24 zeros per second as shown in Figure 1.

6 Very High Frequencies and Ultra High Frequencies

All these waves do different things. For example, 'wireless' waves provide a means of communication, light waves enable us to see things, heat waves make molecules vibrate to cause warming and x-rays can pass through the body and land on film, allowing us to take pictures of our bones.

Figure 1. The Electromagnetic Spectrum.

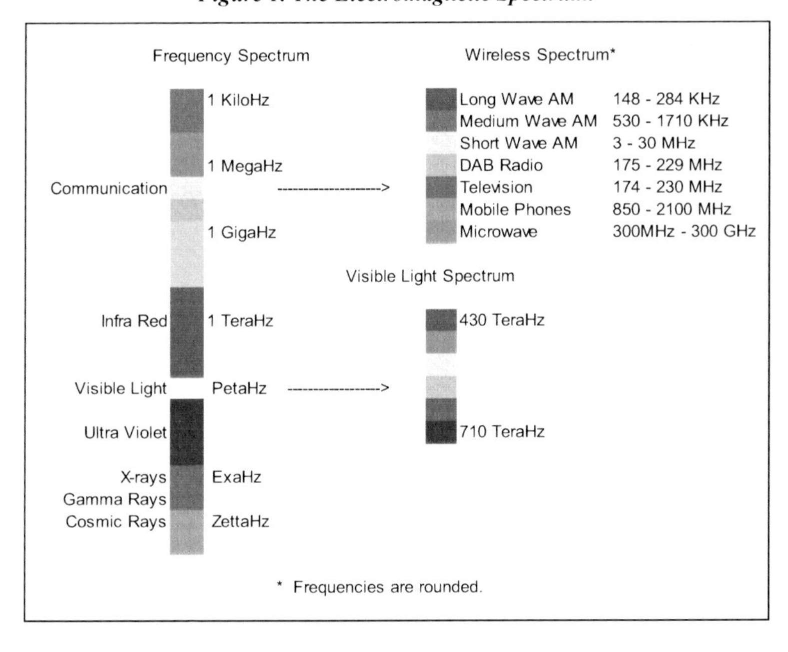

So far, we have talked about the *frequency,* of waves, the speed at which their *amplitude* changes. However, another way to look at them is by their wave*length*. That is, the distance between the crests of the waves. For example, an electromagnetic wave with a frequency of 10,000 hertz has a wavelength of 30 kilometres (18.6miles)[7]. For a frequency of 300×10^{24} hertz[8] the wavelength is 10^{-19} metres[9]. With all this in mind, we will let our imagination run wild for a moment.

---

7 Rounded

8 300 yottahertz

9 0.000,000,000,000,000,000,1 metre

You dream that you are on a beautiful tropical island watching the waves coming ashore. You notice that each new wave follows the one in front (cycles) about every 5 seconds and realise that the waves have a frequency of 0.2 cycles per second. Then you see that the distance between the crests of each wave is about two and a half metres and you know that is their wavelength. With very little effort you work out that the speed of the wave is half a meter per second.

Now you see a piece of wood riding on the surface of the sea, but it does not move towards the shore, or at least only very slowly. Knowingly, you remind yourself that, had it not been for the wind and/or currents, the log would have remained stationery because only the wave energy is moving to the shore not the material. You wake up feeling refreshed and smug.

More About Waves.

When you woke up from your dream you remembered that the atoms on the surface of the waves (those that have not been evaporated by the heat of the sun) are attracted downwards by the atoms in the water below. (Similar to the atoms on the surface of a semi-conductor – see later). This results in a smooth elastic skin on the surface called 'surface tension'.

When the waves (not the water) met the shallow shore they increased in height until the weight of the water overcame the strength of the surface tension and broke up into a myriad of tiny bubbles. When an individual bubble bursts it makes a tiny sound but there were so many of them that they combined to create the crashing sound of the 'breaking' waves.

Radio

We take it for granted today that we can speak to one another from thousands of miles away almost instantly, without any visible connection between the speaker and ourselves. But it was little more than one hundred years ago that the idea was the stuff of science fiction.

The invention of Wireless telegraphy[10] and Wireless telephony[11] can be attributed to many early experimenters, notably those in the USA, Italy, Germany, India and Russia but the foundation was laid in the lecture theatre of the Oxford University Museum of Natural History in 1894. One can imagine the excitement as Professor Oliver Lodge and Alexander Muirhead demonstrated sending a signal without wires; just by Hertzian (radio) waves. During the demonstration, the signal was sent from the nearby Clarendon Laboratory building and received by apparatus in the lecture theatre.

Oliver Lodge

Oliver Lodge was born at Penkhull near Stoke-on-Trent in June 1851 of parents from middle class backgrounds. His father became a commercial representative for a clay pottery company. Oliver left school at the age of 14 but his keen interest in science led to a bachelor's degree at the age of 24 and then his Doctorate of Science two years later. He became a Professor of Physics at University College in Liverpool and was made a Knight and a Fellow of the Royal Society in 1902. He has been described as, "Always an impressive figure, tall and slender with a pleasing voice and charming manner, he enjoyed the affection and respect of a very large circle".[12] But Lodge was another scientist who did not stay the course. His problems were brought about because of his limited expertise in mathematical physics and a reported reputation for being resentful and obstinate. History is fickle.

Oliver Lodge sided with the many scholars who were critical of Marconi's wireless telegraphy that he had invented around the turn of the 19th century. However, before he died, three years after

10 Communication by pulses and codes.

11 Communication by speech, music etc.

12 From an obituary in The Times.

Marconi, his junior, Oliver came to acknowledge some of Marconi's achievements. He even admitted that, at the beginning of his career, he did not appreciate the importance of wireless telegraphy, for the transmission of real messages (telephony) as well as for the transmission of electrical impulses (telegraphy).

Lodge did not know that radio waves travel in a vacuum. Instead, he believed in 'Ether', an imaginary substance assumed to fill the otherwise empty space between the stars and planets[13]. Its presence in space was thought to be necessary to provide a carrier for light waves and radio waves. Lodge clung to the idea of Ether long after the transmission of radio waves through a vacuum was generally accepted. It was probably the idea of 'Ether Drift', the supposed relative motion between the Ether and any body within it that made him a firm believer in psychic phenomena. Consequently, he became a member of The Ghost Club and president of the Society for Psychical Research. This association marred his reputation as a scientist.

Oliver was married to Mary Marshall, of Newcastle-under-Lyme. She bore him a family of six sons and six daughters. He died in August 1940 at Normanton House near Salisbury.

Although Oliver Lodge and Alexander Muirhead *demonstrated* the principle of wireless transmission, it needed Guglielmo Marconi to make a commercial success of it. He is another example of one who benefited by promoting the earlier work of other experimenters and physicists. Born in 1874 as the 1st Marquis of Marconi, he is remembered mainly for his successes in transmitting a signal across the Atlantic Ocean and in attracting Queen Victoria's interest in wireless. Marconi's equipment was installed in the wireless room of the Titanic and contributed to saving many lives. Afterwards, he received an accolade from Britain's Postmaster General who, referring to the *Titanic* disaster,

---

13. We should not be too critical of Oliver Lodge's misconception. Today, scientists believe in the existence of Dark Matter without understanding its mysteries.

said, "Those who have been saved, have been saved through one man, Mr. Marconi ...... and his marvelous invention."

Guglielmo had a close shave. He was offered free passage on the Titanic for its ill-fated maiden voyage in 1910 but instead had boarded the Lusitania three days earlier. It seems that he had taken a shine to a lady public stenographer aboard the ship, which influenced his choice of travel. But Marconi was not the only one to have a narrow escape. Due to the cancellation of a series of concerts in New York, the London Symphony Orchestra also missed sailing on the Titanic.

Marconi pictured in 1896 with an early Spark Gap transmitter and receiver that could send and receive Morse Code.

Like the Titanic, the Lusitania, its sister ship, also met a sad end when it was torpedoed in 1915 with the loss of 1,195 lives.

These early wireless transmitters relied on a series of powerful sparks; pulses of electromagnetic radiation, using the Morse Code.[14] This was fine except that the sparks needed a high and lethal voltage and a very long aerial. Worse, when one powerful transmitter operated, it swamped the others.

The early receivers used a glass bulb filled with metal filings for the detection of the radio signals. When a wireless signal pulse was received, the filings stuck together thereby creating a low resistance path for an electric current. Edouard Branly first noticed the effect but Oliver Lodge improved on the device to make it more suitable for radio reception. He arranged for a clapper that tapped the bulb to unstick the filings, making it ready for the next pulse. Appropriately, the device was called a coherer. One can understand the crying need for a better receiving device.

---

14 A code that used long and short pulses (dashes and dots) named after the American, Samuel F. B. Morse its co-inventor.

Ambrose Fleming came to the rescue. He was a Lancastrian and became the first professor of Electrical Engineering at University College, London. He was also consultant to the Marconi Wireless Telegraph Company, the Swan Company, Ferranti, Edison Telephone, and later the Edison Electric Light Company.

Ambrose Fleming

Fleming's answer for the replacement of the coherer was his invention in 1904 of the two-electrode valve[15], the *diode*. He knew that the hot filament of an ordinary electric light bulb emitted electrons so he inserted a metal plate, connected to a positive voltage, to attract them. Because the electrons could travel only from the filament to the plate (the anode) but not in the other direction, it was a 'one way street' and was called a 'rectifier'.

Fleming's rectifier diode acted like a 'semi-conductor'. (Plate 1).

> Because the glass bulb of the diode is devoid of air, the negative electrons from the heated filament (acting as a 'cathode') are attracted to the plate (called the 'anode') by virtue of its positive potential relative to the cathode.

Wires made of a metal such as copper are *conductors*. Current can flow through them easily. Substances like glass and rubber are *insulators*, little or no current can flow through them. Some elements like Silicon[16] and Germanium are called metalloids and fall between the two. If impurity atoms are added to skillfully contaminate (dope) a pure metalloid, it acts as a *semi*[17]*- conductor*. It behaves in a similar way to a diode valve or a non-return valve in a water pipe; the electricity or water can only flow one-way.

---

15 The word 'valve' is used as in the vernacular of the time. The American usages, 'thermionic vacuum tube', vacuum tube' or just 'tube' are synonymous terms and may be found elsewhere in the book.

16 Not to be confused with Silicone.

17 Literally, 'half'.

But, we are getting ahead of ourselves. We shall see later that the thermionic diode was the seed that blossomed into the semiconductor *transistor* over forty years later.

> Silicon is the eighth most common element in the universe by mass, but very rarely occurs as the pure free element in nature. It is most widely distributed in dust, sands, planetoids, and planets as various forms of silicon dioxide (silica) or silicates. Over 90% of the Earth's crust comprises silicate minerals, making silicon the second most abundant element in the earth's crust after oxygen (about 28% by mass).
>
> *'Silicon'. Wikipedia*

Just three years after the invention of the *diode valve* with its two electrodes (the *filament* and the *plate*) in one glass bulb, Lee De Forest made a huge step forward.

Lee de Forest

De Forest, an America, born in 1873, improved on Fleming's diode by adding a grid arrangement between the filament and the plate. He called his device an Audion, now known as a *triode valve* (three electrodes in one glass bulb - the filament, the plate and the *grid*). This gave it the ability to amplify a weak and controllable signal. His 1906 invention was granted a US Patent in February 1908.

A triode valve is like the water pipe with its non-return valve but with the addition of a tap. Just as a small effort is required to turn a tap to regulate (modulate) a required flow of water, so can a small voltage on the grid of a triode modulate a large flow of electric current. Both act as amplifiers.

Lee De Forest had a troubled life. Not only did he not know how his device worked but also he was involved in several patent lawsuits and spent a fortune from his inventions on legal bills. He had four marriages and several failed companies and was defrauded by business partners. As if that was not enough he was once indicted for mail fraud, although later acquitted. However, he can rest in peace knowing that he gave us the triode, vital in the development of long-distance communications such as radio, television and radar. His triode was an important milestone in the progress of electronics during the first half of the 20th century, culminating with its replacement - the transistor.

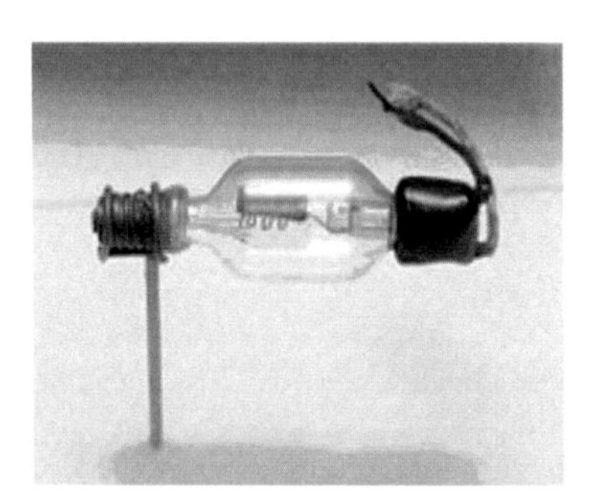

The De Forest Audion of 1906

During the 1914/18 war, radio (wireless) and radio valves, like other technologies, improved by leaps and bounds so that by 1920 wireless programmes were transmitted for public entertainment.

Hundreds of radio stations began broadcasting in American in 1920 but without some form of control they swamped the airwaves. The British approach was more cautious, probably as a result of the American experience. The first tentative experimental transmissions took place at the Marconi Works in Chelmsford, Essex, in 1920. They developed into the first British station to make regular entertainment broadcasts. These started in February 1922 from an ex-Army hut next to the Marconi laboratories at Writtle.[18] Initially the station broadcast every Tuesday from 8.0pm to 8.30pm with the call sign 2MT. (Plate 2). This was followed by to the creation of second Marconi station with the call sign, 2LO (Plates 3 & 4) and subsequently, the establishment of the BBC.

---

18 Near Chelmsford in Essex, England

2LO's temporary base was at the company's head office, Marconi House, in the Strand, London. The transmitter was located in an attic room and the aerials were strung between towers on the roof. Then, at the end of 1922, John Reith was appointed General Manager of the BBC and took up residence in vacant space in the northwest parts of Savoy Hill House in Savoy Place, the premises of The Institution of Electrical Engineers[19]. It housed small studios for the transmission of entertainment content.

Crystal sets, and later battery powered radio sets, were used for the reception of these broadcasts and they soon became popular with the public. Many hobbyists built their own sets using instructions found in magazines. The crystal sets used a 'cat's whisker' detector made with a crystal of Galena and a thin wire contact. It acted as a semiconductor.[20] The cat's whisker was very sensitive to the pressure of the wire-to-crystal contact and the slightest vibration could dislodge it. Long-suffering users needed to move the wire across the crystal surface until he or she was lucky enough to find a 'sweet spot' and hear a radio station or the noise of "static" in a pair of headphones. Then the slightest jog would move the whisker and reception would be lost.

Crystal set receivers could pick up only strong signals and could only drive headphones. They also suffered by being unselective. It was not possible to completely separate adjacent broadcasting stations. This was also a feature of the early battery valve radios. (Plate 5). The problem was solved later with the advent of the superheterodyne[21] receiver. (Plate 6).

> Galena is a natural mineral form of lead sulphide and is a semiconductor. It can act therefore as a diode; albeit a poor one. Due to the impurities in its crude form, only certain places on the surface act as reasonable rectifying junctions.

---

19 Now renamed, The Institution of Engineering and Technology.

20 The name 'cat's whisker' was in common usage because it described the thin wire used to touch the surface of the crystal.

21 Superhet. (col.)

Battery operated radios needed a heavy 2 volt lead/acid 'accumulator', a large, typically 90 volt, battery comprising numerous 1½ volt carbon cells[22] and a 9v 'grid bias' battery of cells. The accumulator had to be taken to the local radio shop for a recharge and the batteries replaced on a regular basis. Mains electricity-powered sets introduced in the early 1930's removed the need for this chore. These mains receivers are remembered by their need to 'warm up' after switching on the set. The 5-valve, mains-powered, superhet receiver was the status symbol of the early 1930's. Most domestic wireless sets provided three wavebands selected by a switch; Longwave, Mediumwave, and Shortwave amplitude-modulated transmissions. (See Figure 1).

Wireless has been used for military purposes from the First World War. The German airships for example used it for direction finding. The equipment at the time was very heavy and limited its use in aircraft because it reduced the weight of the armaments that could be carried. Valve electronics held back progress in the design of radios and public address (PA) systems, until the arrival of the transistor.

Sound Recording

What was it that compelled experimenters to find a way to record and play back sound? Was it ego? Was it to hear their performances for the subsequent analysis? Was it for time shifting? Or did they realise the commercial significance of selling multiple copies of recordings? Whatever the reason, experiments in sound recording took place as far ago as the middle of the nineteenth century.

Edouard-Leon Scott de Martinville made a sound recording machine in 1857. His device created a waveform image of sound on a rotated cylinder of soot-coated paper. He called it a

22 These days, cells (for example AA, AAA and C type) are often called batteries. A battery is really a group of cells.

phonautograph. But his machine created only visual images of the sound and could not play back the recordings.

The Valve

The filament of the triode (acting as a 'cathode') emits electrons and these are attracted by a positive potential on the plate (anode). On their way they have to pass through a grid. The size of the voltage applied to the grid determines how many electrons can get through to reach the anode.

For battery operation, the filaments of valves (the source of the electrons) were heated by direct current. If alternating current, derived from the mains, was applied to the filament instead, the valve would generate a 50 Hz hum noise that would swamp the genuine signal. As a result, the filament (now aptly called the heater) was made to heat a metal sheath (the cathode), which emitted an even flow of electrons. It was this process of waiting for the cathode to heat up that caused the delay in start-up. See diagram below.

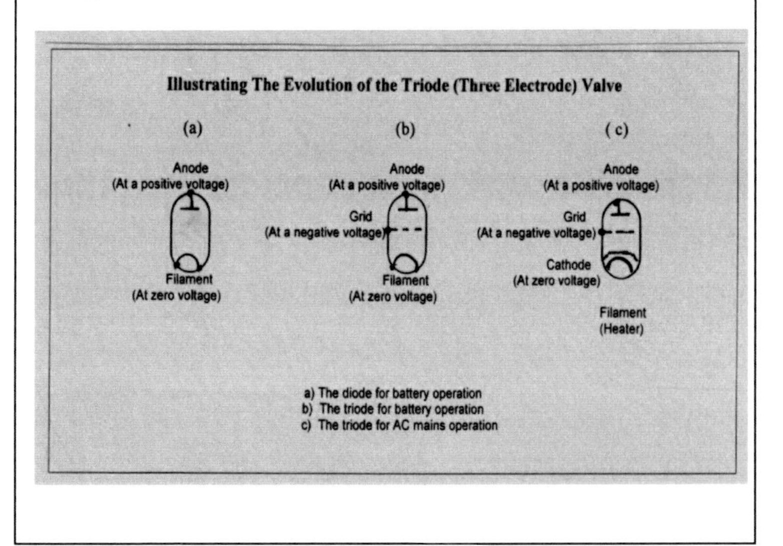

Early Rectifiers

The years preceding World War I saw progress in the use of lead sulphide (galena) and lead selenide materials for rectification purposes. They were used in equipment for detecting ships and aircraft, for infrared rangefinders and for voice communication systems. Later, the point-contact crystal detector became vital for radar and microwave communication because the valves that were available could not respond as detectors above about 4000 MHz. Therefore, considerable research and development of silicon materials occurred during the war to develop detectors of consistent quality. From the 1920's, to World War II in the 1940's, selenium and copper oxide rectifiers and germanium semiconductor diodes were used extensively for power supplies and for signal rectification as alternatives to vacuum tubes. Later, transistor technology paved the way for efficient high power silicon rectifiers.

Emile Berliner

Then, in 1887, Emile Berliner, a German-born American inventor, took a great step forward. He fixed Scott de Martinville's sooty images with varnish and photoengraved a corresponding groove into the surface of a metal cylinder. This, for the first time, enabled recordings to be played back as sound.

Emile Berliner was born on the 29th May 1851. During his lifetime, he overcame the problem of mass production of recordings. He conceived the idea of forming the grooves on a flat disk rather than on a cylinder and patented his idea of a gramophone in 1887. His disk, that he called a 'gramophone record, used lateral movement of the needle (stylus) meaning that it moved from left to right in accordance with the sounds instead of up and down. Emile founded the Berliner Gramophone Company in 1895, and recorded music as early as 1894. He died in August 1929.

The audio fidelity of these early gramophone disk recordings was about the same as the phonograph cylinders but the relative ease by which discs could be produced in quantity led to the demise of the cylinder. These and other systems at the time were acoustic only. It was too early for electronics to play a part.

Thomas Edison

The American, Thomas Alva Edison, is one of the most famous inventors. Not only a determined, prolific inventor, he also developed into an industrialist and business manager. This probably accounts for his standing in history. Many other great inventors were content to stay in the 'back room' and therefore remain unrecognised.

Born in February 1847, Thomas was not successful at school. He left after three months and his mother taught him at home. Clearly, she must have done a good job but his lack of academic study would explain why he had a tendency to speak with disdain for academia and corporate operations.

Edison was fascinated by the idea of recording and playing back sound. He perfected the phonograph in 1878, a device with a cylinder covered with an impressionable material such as tin foil, lead or wax on which a stylus etched a track with vertical grooves. A needle passing through the grooves played back the recorded sound and was amplified mechanically.

Copies of these cylinders were limited to about twenty-five, all of significantly lower quality than the original and destroying the original in the process. Even by using of a number of machines to record multiple originals, a single performance could produce only a few hundred copies for sale. By 1902, successful molding processes for cylinder recordings were developed.

Thomas Edison

Edison invented the incandescent light bulb and formed a lighting company. In 1882 he built a power station in Pearl Street, Manhattan that generated enough electrical power using direct current (DC) at 110 volts to serve 59 customers. He also developed the motion picture camera and the alkaline storage battery. While in his eighties. Edison became friendly with Henry Ford and worked on several projects. These ranged from electric trains to finding a domestic use for natural rubber.

Nikola Tesla, an academically qualified engineer, once worked with Edison's company. However, they later became longstanding rivals. They publicly clashed about the use of direct current (DC) electricity, favoured by Edison, versus alternating current (AC). A business dispute developed over the choice. Apparently, Edison tried to convince people of the dangers of alternating current by using public demonstrations in which animals were electrocuted. One of the most infamous of these was the electrocution of a circus elephant in New York's Coney Island in 1903. We now accept that, for a like-for-like voltage, AC is safer than DC.

Thomas Edison died in 1931.

Electrical Recording

The advent of electrical recording improved the quality of the recording process of disk records considerably. From 1925 to 1930, the foremost technology for home sound reproduction comprised a combination of electrically recorded records with spring-driven acoustic gramophones with folded horns. They gave a reasonable quality of sound.

Later, the speed at which the disks rotated was standardized at 78 revolutions per minute (rpm). Further innovation allowed speeds of 45, 33⅓ and even as slow as 16 rpm for broadcast transcriptions and some rare consumer recordings. Electronic amplification, using valves, replaced acoustic systems from 1925 and vinyl began to replace shellac for the disk material in 1948 with the advent of LPs.

Electrical recording made a number of other new features possible. For example, it became more feasible to record one track to disk and then play that back while another track was played with both tracks being recorded onto a second disc. This, and conceptually related techniques, were known as *over-dubbing*.

The advent of analogue audio magnetic tape and over-dubbing, pioneered by Les Paul, called 'sound on sound' recording, was the forerunner of multi track systems.

The electrically recorded shellac, 78 rpm, records used up-and-down ('hill and dale)' vertical movement of the needle (the stylus). The preference for lateral movement returned when 33⅓ and 45 rpm vinyl discs replaced them. Lateral movement was essential for recording stereo.

Electrically powered gramophones were introduced around 1930, but crystal pick-ups and electronic playback (using valves) did not become common until the late 1930s. A floor-standing cabinet containing both a radio and a gramophone, appropriately called a 'radiogram', became another example of a status symbol of the period.

Between 1925 and 1930, record companies switched to the use of electric microphones to capture the sound. They greatly improved the sound quality. However, presentations were still impressed directly on to the recording medium so, with live performances, if a mistake was made, there was no way back.

Valdemar Poulsen

Valdemar Poulsen was a remarkable man. Born in 1869, in Copenhagen, Denmark, he invented and outlined the basics of many of the recording devices that are commonplace today. He is regarded as the founding father of the audiocassette, the CD and DVD, the computer disk and diskette, the credit card, the iPod and just about every other piece of equipment today that records sound or data.

Poulsen demonstrated magnetic recording in principle as early as 1898. His magnetic recorder used a steel wire moving at high speed past a recording head. The sound, in the form of an electrical signal, was fed to the recording head, resulting in a pattern of magnetization in the wire. Playback was by means of a similar head that picked up the changes in magnetic field from the wire and converted them back into an electrical signal. His apparatus was limited because valves were not yet invented to amplify the signal.

At the1900 World's Fair in Paris, Poulsen had the chance to record the voice of Emperor Franz Josef of Austria. It is believed to be the oldest surviving magnetic audio recording.

Over twenty years later, in the 1920's, Curt Stille, a German inventor, developed the telegraphone. It evolved to become a commercial wire recorder but the reproduction quality was significantly lower than that achievable with a phonograph disk. Nevertheless, wire machines were used to record dictation during the 1940s and early 1950s. They suffered from a tendency for the wire to become tangled. Splicing was possible by tying together cut ends of the wire but it was not satisfactory remedy.

On Christmas Day, 1932 the British Broadcasting Corporation first employed a potentially lethal Marconi-Stille system. It used 3 mm (0.1") wide and 0.08 mm (0.003") thick steel razor tape running past the recording and reproducing heads at 90 metres (approximately 300 feet) per minute. The length of tape required for a half-hour programme was nearly 3 kilometres (1.9 miles) and a full reel weighed 25 kg (55 pounds). It is surprising that John Reith allowed it.

Fortunately, magnetic tape replaced the use of steel wire, so avoiding the attentions of the Health and Safety inspector. One hopes that the poor man operating the Marconi-Stille recorder, shown in Plate 7, survived to use magnetic tape.

German engineers at AEG, working with the chemical giant I G Farben, demonstrated the world's first practical magnetic tape recorder in 1935. They called it the 'K1'.

The first stereo recordings on disks were made in the 1930s but they were never issued commercially. Then around 1943, German audio engineers developed multitrack recording, in which a magnetic tape was divided into adjacent tracks. Stereo sound used left and right channels on separate tracks. It was adopted for modern music in the 1950s because it enabled signals from two or more separate microphones to be recorded simultaneously, so that stereophonic recordings could be made and edited conveniently.[23] Stereo sound gave a considerable improvement to the realism of the recording.

Biasing

During World War II, an engineer at the Reichs-Rundfunk-Geselschaft discovered the effect of adding an AC bias to the recorded signal. It was an inaudible high-frequency signal, typically in the range of 50 to 150 kHz. The biasing process radically improved the sound quality of magnetic tape recordings so that they became comparable with live programmes.

Valve-based magnetic tape, 'reel-to-reel recorders for the home were developed in the late 1940s and early 1950s and became popular with the public. They used 7-inch diameter reels of quarter inch wide tape running at 7½ inches per second. However, they were heavy and bulky. Small, portable products such as the Sony 'Walkman' had to await the arrival of the Compact Cassette and the benefits of the transistor and its progenies.

23 Wikipedia. 'History of multitrack recording'. 8th August 2014.

Early Magnetic Tape Recording

Immediately after World War II John Mullin, an American audio engineer, was given two suitcase-sized AEG 'Magnetophon' high-fidelity magnetic tape recorders. Over the next two years, he modified and improved their performance and gave two public demonstrations of his machines. They caused a sensation among American audio professionals. Many listeners could not believe that what they heard was not a live performance.[24]

Television

We are fortunate that our eyes have persistence of vision.

Persistence of vision is the commonly accepted term for the phi phenomenon. It causes images on the retina of the eye to persist for approximately one twenty-fifth of a second. So, if a series of still images are shown rapidly in sequence, we experience the illusion of an image in continuous motion. Without this trompe-l'œil there would be no cinema, no television, no camcorders or flipbooks. Huge industries would never have existed.

Flicker

In the early days of the cinema it was found that a frame rate (the speed of a series of consecutive images) of less than 16 frames per second caused the mind to see flashing images, or 'flicker'. We can still interpret motion at rates as low as ten frames per second or slower but the flicker caused by the shutter of a film projector becomes distracting below the 16-frame threshold. This distraction is apparent with film, camcorders and television.

A practical form of television requires three essential features beyond those of a cinema film.

---

24 Thanks to Peter Hammer. MIX 1999

First, television is dependent on a device (a special camera) that can rapidly turn light intensity into electricity; it needed a *photoelectric device.* Second, it is necessary to *scan* each image frame in order to break up the picture into tiny segments[25] and transmit these rapidly one after the other. And finally, the receiver must be in perfect *synchronism* with the transmitter to reconstitute the image frame. The inventors of practical television systems had to find the answers to these three requirements, the most difficult being the last.

Alexandre Becquerel

In 1839 Alexandre Edmond Becquerel, a Parisian physicist, observed that a voltage was generated between a solid and a liquid electrolyte when struck by light. So, when only nineteen years old, he had discovered the photoelectric effect; the *first requirement* required for television. Having a doctorate from the University of Paris, Becquerel took a professorship at the Agronomic Institute of Versailles. His name was immortalized when it was used for the SI derived unit of radioactivity, called the 'becquerel' with the symbol Bq. He died aged 52.

Willoughby Smith

In 1873, Willoughby Smith developed a method for continually testing an underwater cable as it was being laid. For his test circuit, he needed a good semi-conducting material[26]. He selected selenium rods for this purpose. The selenium appeared to be satisfactory in the laboratory but in actual use, it gave inconsistent results. It was then discovered that resistance of the selenium rods decreased significantly when exposed to strong light. Smith described the effect in an article in 'Nature', published on the 20th February 1873.[27] So, now, the photoelectric device, called a Selenium Cell, was available. It was one of the earliest known semiconductor materials and was another discovery that paved the way to the transistor.

---

25 Strictly speaking, this is more descriptive of a digital television picture. However the principle is the same.

26 A material that acts like a diode; a one-way street' to the passage of electric current.

27 The "Effect of Light on Selenium During the Passage of an Electric Current"

Jan Van Szczepanik has been described as 'a brain the size of a planet'. He was born in the Austrian Partition in 1872 and became a prolific inventor ranging from photographic systems, a duplex rotor helicopter, a dirigible, a submarine and developments in moving wing aircraft. But most relevant to our story is that he was one of the early pioneers of television development.

Jan Van Szczepanik

He addressed the *second requirement* required for television by inventing a method for scanning. His system was based on vibrating mirrors as shown in Figure 2.[28] However, although Szczepanik used a photoelectric device, (the first requirement), and devised and an ingenious scanning system, (the second requirement), he never managed to solve the third, *synchronization.*

Figure 2. Jan Van Szczepanik's Apparatus.

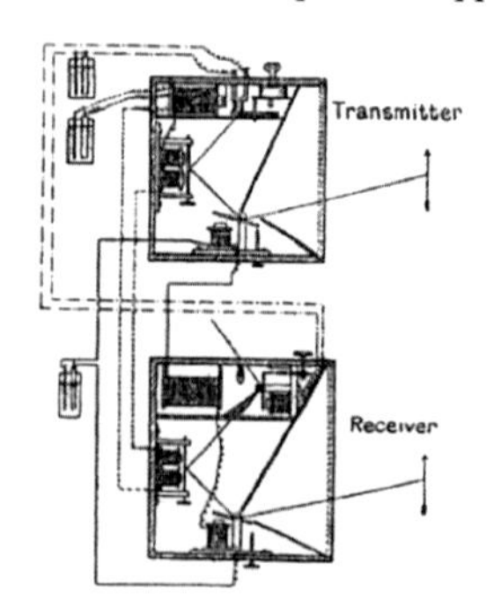

Jan Van Szczepanik died in 1926 at the age of 53, the year that John Logie Baird demonstrated what is believed to be the first true television ever witnessed.

From the time of Szczepanik's apparatus at the turn of the 19th century to Logie Baird's demonstrations in 1924, there were

28 From 'A History of Early Television'. Volume 1. 2004. Stephen Herbert. ISBN 0-415-32666-4

several attempts to devise a viable means of transmitting moving images but *synchronization* remained the stumbling block. In the interim, various methods were proposed for scanning.

Lee De Forest's invention of the first triode valve in 1907, and the work of German, Arthur Korn, and others, made the idea of television a practical proposition. The triode provided amplification for the photoelectric cell and other functions and it gave a means for wireless transmission and reception of the television signal. But, crucially, it helped to overcome the problem of synchronization.

Logie Baird

John Logie Baird was born in Helensburgh on the west coast of Scotland. As an engineer and inventor he is remembered as 'The Father of Television' He evolved a practical television system that included a method of synchronization of the transmitted and received signal - *the third requirement.*

A German book on the photoelectric properties of Selenium claimed Baird's interest in television as early as 1903. Then in September 1920, after a brief spell in business in London, he began to devote his life to experimenting with television.

His first achievement took place in Hastings in 1924 when he transmitted the image of a Maltese cross over the distance of 10 feet, followed by a moving image (Plate 8). The big breakthrough came in October 1925 when Baird achieved television pictures with light and shade (half-tones), making them much clearer. He demonstrated these to invited members of the Royal Institution in January 1926 with images measuring 3.5 x 2 inches.

In April 1925, Gordon Selfridge invited Baird to give personal demonstrations of his television for three weeks at his London, Oxford Street store and paid him £25.00[29] a week.

Baird was dogged with ill health throughout his life, yet his inventiveness and energy were without bounds. He sent television pictures from London to New York by short-wave radio in 1928, demonstrated primitive 'high-definition' colour and 3D television, devised a system for sending images very rapidly (an early fax) and developed a video recording system, that he called 'phonovision'. His company introduced the world's first mass produced television set in 1930 named 'The Televisor' (Plate 9). It gave only crude moving images but they were revolutionary at the time.

John Reith, the BBC's dour General Manager, had disliked television ever since the appearance of Baird's work in 1926. Apparently, as fellow Scots, the two men had crossed swords in their student days at their technical college in Glasgow. However, reluctantly, he came to an arrangement with Baird to send out experimental television transmissions in 1929 provided initially Baird paid the BBC for the privilege. Then, although the BBC began using Baird's system for the first public television service in 1932, the subsequent history is a sad one.

In 1934, the British government set up a committee (the "Television Committee") to advise on the future of television broadcasting. The committee recommended that a high definition service (defined by them as being a system of 240 lines or more) should be established and run by the BBC. Its recommendation was accepted and tenders were sought from industry.

Two tenders were received, one from the Baird company, offering a 240-line, mainly mechanical system, and the other, a 405-line, fully electronic system from Marconi-EMI.

---

29 Worth about £1,400 in 2016 money value

Perhaps surprisingly, the Television Committee advised that they were unable to choose between the two systems so both tenders were accepted and the two ran alternately for an experimental period. So, an embryonic BBC television service began to broadcast from studios in Alexander Palace in November 1936. Both systems relied on the use of valves.

In the event, Baird's 240-line system proved to be inferior to that of Marconi-EMI. Worse, it was crude. His system required that, first, the live program had to be filmed with a normal cine camera. Then the film was passed through a developing bath, dried and run past a flying spot scanner to convert the image into a television signal. Not only was this costly to run and imposed a time delay in the transmission process but the developing bath used sodium cyanide, an odorous and highly toxic chemical. Clearly not a process designed to win the favour of the operators and performers.

In January 1937 after three months of trials, the Baird system was abandoned in favour of the 405-line Marconi-EMI system; a system that became the standard for all British TV broadcasts until the 1960s. The signals were transmitted on a carrier frequency of 45 MHz. (6.7 metres).

There is some belief in the view that the Government had a say in the choice of the system because it had an interest in the development, and availability, of cathode ray tubes and special valves for radar development in preparation for an anticipated conflict with Germany.

It was some consolation that Baird's achievements were recognised later when the Royal Society of Edinburgh awarded him an Honorary Fellowship in1937.

The BBC television service ceased transmissions on the first of September 1939, three days before the outbreak of World War II, The fear was that the transmissions would aid the navigation of enemy aircraft. Also, every available television engineer was required for work on radar. The service recommenced in 1946.

Then, as with the early days of radio, hobbyists built their own 405-line television receivers. Some used kits but the more adventurous bought modules and components, such as the ubiquitous EF50 valves, from the many Government war surplus shops and designed their own circuits (Plates 11 & 12). The writer remembers building his first television set using a 2½-inch radar cathode ray tube with a green screen and the thrill of viewing the 1948 Olympic Games.

By the 1950's, black and white television used around 20 valves. The fact that the valve's heaters alone consumed some 100 watts[30] indicated that, like radio, progress was severely limited until the transistor became available.

Video Recording

We have seen that John Logie Baird demonstrated his 'phonovision' video recording system in the late 1920's, which in the context of the time was imaginative. But this idea of recording the vision signal on a normal gramophone record was not really a practical proposition. It resulted in a barely recognisable 30-line image. Since then, the challenge of finding a way to record a 405-line (or greater) picture occupied engineers for over three decades. It was not until the transistor became availability that a viable commercial solution emerged.

---

30 Some three times more than a modern flat screen colour television consumes.

Television Scanning

The BBC/EMI television scanning system comprised a series of horizontal lines. Each line scanned from left to right with a rapid return to the left, ready for the next line below to begin. The greater the number of lines per image-frame the higher was the resolution and therefore the quality of the picture. Baird's original mechanical system had only 30 lines running vertically. For the BBC trials, his system had 240-lines while EMI used 405-lines for horizontal scanning with synchronising pulses added to the ends of the lines and image frames.

The picture below shows a scan of just 12 lines for clarity. The thin dotted lines are the visible picture paths of the scan and the solid and thick dotted lines are the invisible, rapid 'flyback', return paths, of the beam. For simplicity, the diagram takes no account of 'interlacing' of alternate lines, a feature used by the BBC transmitted system.

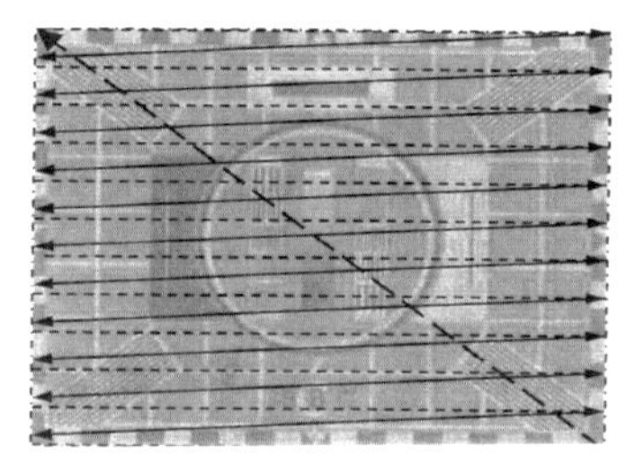

Computers

Although our bodies have evolved to an advanced state, we have always strived to improve on its design. But what are the shortcomings that make the quest worthwhile?

We, as human 'computers', have hormones like testosterone and estrogen. These hormones may be useful for producing 'minicomputers', albeit very demanding ones and with high running costs, but they introduce unpredictability.

We have the ability to do the three R's but we were not very fast or very accurate at doing them. We get bored and careless if we have to repeat the same things over and over again. However we are quite good at picking things up and running around.

Also, with the passing of time, we have become more demanding and more expensive to run. Increasing affluence has made the cost of employing us to do simple repetitive tasks unacceptably high.

Mechanical machines to calculate numbers, such as the abacus, appeared over 4,000 years ago. Since then, even as late as the 1930s/1940s, scientists, engineers and commerce were limited to the use of various mechanical devices such as the slide rule, the Brunsviga 10 stepped gear calculator (Plate 10) and the typewriter.

Within our skull, we have a central processor, a memory, and a means of data input and output, together called a brain. Put simply, the processor controls the memory and the various parts of the body.[31]

The brain is hidden from sight and we are only aware of its existence because of its affect on the 'peripherals' that are connected to it; our eyes, ears, arms, legs and things[32] (and if we have a headache). Furthermore, we are aware that we have a brain because of consciousness, which puts us one up on the man-made computer - for the present at least.

Charles Babbage

Charles Babbage is known as the 'father of the computer'. Born in 1791, he became outstanding among the many polymaths[33] of the 19th century. He is said to have invented the first mechanical computer. This seems a little unfair on Joseph-Marie Jacquard who invented a loom in 1801 that used a punched

31 For our purposes, we assume that these parts of the brain are separate entities. For readers who want a more accurate description of how the brain really works, I recommend a PhD in physiology and a lifetime devotion to brain research.

32 In the computer world, peripherals are things like keyboards, mice, printers, scanners, loudspeakers etc.

33 One with and encyclopedic knowledge and expertise that spans a significant number of different subject areas.

card system.[34] Jacquard's machines were the first step toward automated weaving and had an influence on later scientists, including those working on the digital computer. Nevertheless, Babbage is the one who is remembered for originating the concept of a programmable system.

When he arrived at Trinity College, Cambridge, in October 1810, Babbage was already self-taught in the mathematics of his time. As a result, he was disappointed with the standard of mathematical instruction he found there. Twelve years later he proposed the design of a *difference engine* using the decimal numbering system. It was powered by cranking a large handle. Babbage submitted a paper to the Royal Astronomical Society on the use of such a machine entitled, "Note on the application of machinery to the computation of astronomical and mathematical tables".

The British Government had an interest in Babbage's proposed machine, because producing tables of numbers by hand was time-consuming and expensive. They hoped the difference engine would make the task more economical. Consequently, in 1823, Babbage received £1700[35] to start work on the project. Although his design was technically feasible, it proved to be much more difficult and expensive to complete than anticipated. By 1834 the mounting costs stopped his attempt at the construction of the difference engine by which time the government had put £17,000 into the project and Babbage had contributed £6,000 of his own money.

The Difference Engine

A difference engine was an automatic mechanical machine designed to give error-free tabulation of polynomial functions. It could compute many useful tables of numbers for the benefit of engineers, scientists and navigators.

---

34 Also unfair to the Abacus (circa 2500 BC) and Orrery (Circa 150 BC)
35 Appoximately £140,000 in 2016 money value.

It is recorded that Robert Peel's government procrastinated for eight years from 1834 to 1842 before making the decision not to proceed. Dubbey [36] writes:

> "Babbage had every reason to feel aggrieved about his treatment by successive governments. They had failed to understand the immense possibilities of his work, ignored the advice of the most reputable scientists and engineers, procrastinated for eight years before reaching a decision about the difference engine, misunderstood his motives and the sacrifices he had made, and ... failed to protect him from public slander and ridicule."

Babbage was never able to complete construction of any of his machines due to conflicts with his chief engineer and lack of adequate funds. We had to wait until the 1940s for the first general-purpose electronic computer to be built.

Charles Babbage died on 18 October 1871. He appears to have been an 'ideas man' for machines, the construction of which were far in advance of his time.

Babbage's Proposed Computer

Babbage named five logical components of his machine; the mill (the processor), the store (the memory), the control, the input and the output. The store held 1000 numbers each of 50 digits, but he designed the analytical engine to effectively have infinite storage. This was done by outputting data to punched cards that could be read in again at a later stage when needed.

One hundred and fifty three years after it was designed, a complete Babbage Difference Engine No. 2 was completed by the Science Museum in London in 2002. Built faithfully to the original drawings it comprises 8,000 parts, weighs five tons, and measures 11 feet long. (Plate 15).

---

36 John Michael Dubbey. Mathematical Work of Charles Babbage. (Cambridge, MA: Cambridge University Press, 1978)

The rapid calculation of the flight of artillery shells became a pressing need in both World Wars. The computations had to take into account the effects of gravity, the initial velocity of the missile, the density of the air, the altitude, the temperature and the humidity. Human workers could not do the calculations fast enough so much ingenuity went into attempts to ease the problem using *analogue* computation.

World War II stimulated the development of analogue computers of which the American Norden Bomb Sight is an example. (See Plate 17). Then, the urgent need to decode the signals communicated by the German Enigma and Lorenz ciphers, stimulated the development of a digital computer

COLOSSUS

To help break the enemy codes, a semi-programmable computer was built using valves. With the name Colossus, it became the first electronic, stored program, general-purpose *digital* computer. Tommy Flowers at the Post Office Research Station in Dollis Hill, close to London, designed the machines and paved the way for the future of computing in England. The Mark 2 version of Colossus used 2400 radio valves, about 100 logic gates and 10,000 resistors connected by 7 km of wiring. It consumed over 5 kilowatts of electrical power. Crucially, this power was mostly waste heat from the valve's' heaters.

It was believed that the failure rate of valves would make impractical a machine that depended on so many of them but Flowers had discovered that switching them on and off was the chief cause of failure. He knew that his machine would be feasible if it was left on permanently so that the failure rate would be acceptably low.

It was the combination of the exceptional talent of Alan Turing in conceiving the principle of the stored-program, general-purpose computer[37], the competency of Tommy Flowers in designing the electronics and the brilliant decrypting skill of W. T. (Bill) Tutte that enabled the code of Lorenz to be broken at Bletchley Park just in time for D-Day.

Colossus is one of many examples of the technical requirements of war that evolved to benefit society in peace; an example of swords into ploughshares. Although it had the distinction of being the first real, digital computer, those involved were denied any credit because the British Government ordered it to remain a 'secret forever'. This left the field open for others to claim the honour.

Alan Turing

Tommy Flowers

Bill Tutte

Sinclair McKay gives a moving account of the genius of Tommy Flowers and his struggle with politics at Bletchley Park.[38]

ENIAC

An American electronic engineer, J. Presper Eckert Jr, together with physicist John Mauchly built a decimal-based electronic computer in 1945. It was called **E**lectronic **N**umerical **I**ntegrator **A**nd **C**omputer; known as **ENIAC**. The machine's enormous size is indicated overleaf.

---

37 Described in his paper, "On Computable Numbers, with an Application to the Entscheidungsproblem" (submitted on 28 May 1936 and delivered 12 November). Although the idea of allowing computer programmes to be stored in computer memory is usually attributed to John von Neumann's architecture, it appears that he had read Alan Turing's paper. Therefore, the accolade for the invention of the first programmable electronic digital computer must surely remain with Alan Turing.

38 'The Secret Life of Bletchley Park'. Aurum Press Ltd. ISBN 9781781315866

Number of valves:	17,468 (compared with Colossus' 2,500)
Resistors:	About 70,000.
Capacitors:	10,000.
Relays:	1,500
Hand soldered joints:	5 Million.
Weight:	30 tons
Power Consumption:	174 kilowatts.
Size:	10 feet high and 3 feet wide. (The cabinets, which stood side by side in a large 'U' shape, would have stretched for 100 feet if laid end to end).
Input and output:	IBM card readers and punches.

Unlike Colossus, ENIAC (See Plate 16) was switched on and off and thereby suffered a valve failure every two days. It must have been most frustrating for the operators. Perhaps this was an example of Turing's view: ".... in the American tradition of solving one's difficulties by means of much equipment rather than by thought."[39] Nevertheless this was a remarkable feat of engineering because little was known about digital circuitry at that time. One cannot but feel envious of someone actually being paid to carry out such a fascinating and challenging task.

EDVAC

Another American valve-based computer, this time a binary-serial machine called the **E**lectronic **D**iscrete **V**ariable **A**utomatic **C**omputer (**EDVAC**), followed the ENIAC. It used 1,000, 44-bit 'words'[40] to give it an average addition time of 864 microseconds (about 1,160 operations per second)[41] and its average multiplication time was 2,900 microseconds (about 340 operations per second).

---

39 From 'Alan Turing and His Contemporaries"
40 In computer parlance a 'word' is a unit of data made up of a number of 'bits'.
41 Compared with billions of operations per second that is typical of today's Personal Computers.

Analogue and Digital

Before 1950 most electronic engineers had solved their problems using analogue techniques. Many found it difficult to migrate to the techniques of the emerging digital electronics. Yet, in many ways it simplified things and widened the scope of possibilities, and at much lower cost. The digital computer is an example.

So, what is the difference between analogue and digital?

Analogue methods are concerned with continuously varying quantities. In electronics, the quantities are represented by continuously varying voltages or currents. Digital methods are concerned with representing the quantities by groups of just two states, '0' or '1', called 'bits'. The bits are created electronically using switches that are either 'off' or 'on'. The voltage or current 'on' levels remain constant.

For example, the cross section area of a cylindrical container and the height of water it contains, is a measure the volume of the water. Smooth variation in the height of the water is like an analogue signal representing changes in volume.

Now imagine a set of railings with some of them missing. The rails that are present are the '1's and the ones that are missing are the '0's. That is similar to a digital signal.

Apart from lending itself to complex logical processes, digital electronics has a number of other advantages. One in particular is the relative ease of reducing the effects of *unwanted* random electrical signals, known as 'noise'. All electronic components working at temperatures above absolute zero (-273 degrees centigrade) create noise as their molecules dance around. Also, poorly designed electronic equipment is susceptible to interfering electrical noise from external sources. Noise also reaches the earth from space, noticeable as the hiss sometimes heard on radio receivers between stations. Electrical noise can be the cause of an internet connection to drop out or slow down, as the modem attempts to filter it out.

With analogue circuitry, each time the signal is amplified, the noise is amplified with it. The relative magnitude of the signal and the noise is called the 'signal to noise ratio'. This can become so small it may be impossible to distinguish the program content of a radio broadcast from the background noise.

Noise also adds extra random information to digital signals. However, this noise is usually much lower in amplitude compared with the 'on' states of a digital signal. This, together with software filtering techniques, can make digital electronics immune from the effects of the noise; one advantage of the digital transmissions of television and radio.

EDVAC used an ultrasonic serial memory to give automatic addition, subtraction, multiplication, programmed division and automatic checking.

With a weight of 17,300 lb (7,850 kg) it needed eight-hour shifts using thirty operating personnel.

EDSAC

The **E**lectronic **D**elay **S**torage **A**utomatic **C**alculator (**EDSAC**) followed EDVAC immediately after the Second World War. It was built originally in the Cambridge University Mathematical Laboratory in England by a team led by the late Professor Sir Maurice Wilkes. The hardware was over two metres high and occupied a ground area of 215 square feet.

Earlier machines were dedicated to a single task. For example, EDVAC for ballistics research, Colossus for code breaking and the Manchester University "Baby" Small Scale Experimental Machine, for purely experimental work. However, it is generally accepted that EDSAC was the first practical *general purpose* stored program electronic computer. Also, it was the first design to be used commercially in the UK. Based on EDSAC, a computer called LEO-1 was built for J Lyons for scheduling deliveries to its shops.

EDSAC

EDSAC's main memory comprised 1024 words, though only 512 were available initially. Each word contained 18 bits, but the first bit was unavailable due to timing restrictions, so only 17 bits were used. An instruction consisted of a five-bit instruction code represented by a mnemonic letter, so that the *Add* instruction, for example, used the bit pattern for the letter *A*.

This illustrates the awesome progress from these early machines to the personal computer of today.

For some time, punched tape and punched cards served as long-term memory. Hard Discs, USB's CD's, DVD's and even floppy discs were no more than a twinkle in the eye of the future.

Low cost and fast Random Access Memory (RAM) was the Holy Grail and various methods were devised to find it. The use of cathode ray tubes, delay lines, and magnetic cores were just some attempts but it was not until the transistor arrived that it became a viable solution.

Early programming was limited by the low speed of the processors and the small size of the memories, so programmers had to use their ingenuity to get the best they could with very limited hardware and mainly binary notation. Their achievements are surprising and commendable.

I had the privilege of employing Sheila Quinn, a past member of the EDSAC team. She typified the high calibre of those who worked on the project and from her I learned much about their skill and ingenuity at that time.

Signs of The Times

During the 1970's, the late Sheila Quinn, who was one of the country's most experienced software designers, attended a tender evaluation meeting at a West Country water company. I learned afterwards that our company did not win the contract because the person to be in charge of the software development was to be a woman (Miss Quinn). Apparently, the main objector on the committee had been a woman!

Ann Dowling FRS was made President of the Royal Academy of Engineering in 1995 and Naomi Climer, was elected President of the Institution of Engineering and Technology in 2015.

The native language of a digital computer is binary. Its vocabulary is limited to just two numbers, '0' and '1', and this was the only way for users to communicate with the early machines. Figure 3 shows the lines of 0's and 1's a user had to type in order just to enter the word 'Hello'. On can imagine the tedium of typing a letter or an email that way today.

Figure 3. An Example of Binary Code Programming

000010	**Start of text**
001101	**Carriage return (New line)**
100100	**H**
110011	**E**
110110	**L**
110110	**L**
111011	**O**
000011	**End of text**

As with all new technology, standardisation was non-existant so no two machines were alike. Each machine would have its own method for doing the same task.

Complex operating systems and languages were needed that allowed humans to interact more easily and rapidly with the computer. Operating Systems with a **G**raphical **U**ser **I**nterface (**GUI**), like Microsoft's ‘Windows’ or Apple's ‘Mac OS X Lion’, and languages such as 'C' and Visual Basic were yet to come. They had to await the arrival of the transistor and its derivatives to provide large, inexpensive solid-state random access memories and permanent storage systems such as the Floppy Disk Drive, the Hard Disk Drive (HDD) and the Solid State Drive (SSD).

Figure 4. A Historic Comparison of Computers.

	1943 COLOSSUS	**1945 ENIAC**	**1949 EDVAC**	**2015 PC**
Memory/size	Tape	Tape Reader and punch	5.5 Kilobytes)	3 Terabytes[42]
Number of Valves/ Transistors	2,500 valves	17,468 valves	6,000 valves	> 1 Billion transistors
Power Consumption	8 kW	150-200 kW	56 kW	<100 watts
Speed:	5,000 Characters per second. (27.3 mph.)	5,000 Hz	340-1160 Hz	>3 GHz[43]
Size (footprint)	187 sq. ft.	1,800 sq. ft.	490 sq.ft.	< 2 sq.ft.
Cost at the time.	Unknown but very high	Unknown but very high	$500,000	~ £300.00

In a world where calculating machines were widely used and understood, it took many decades before the public felt at ease with the concept and adaptability of the general purpose, programmable, digital computer. During that time the computer industry and

42 1,000,000,000,000. or (10^{12}) bytes
43 >3,000,000,000 Hz

computer aficionados (often called 'nerds' and known for their duffle coats) had a field day in a market of fear and ignorance.

Even as late as the 1970's, most people including members of the Government, with experience limited to the use of desk calculators, disbelieved that that computers had anything to do with the alphabet and human language. In earlier times, they may have been right but the advent of the transistor made sure they were wrong.

Transport

Picturesque railway trackside signal boxes are now few and far between. With few exceptions, gone are the days when a signalman would arrive each morning, and mount the wooden steps to spend the day in his cozy 'office' with a coal fire and an all-round view of his daily responsibility.

From the days of Stevenson's 'Rocket' locomotive of 1829, control and communication on the railways were almost entirely by mechanical and simple electrical and electromechanical systems. Huge, heavy levers changed track points and signals, level crossing gates were raised and lowered manually by huge wheels, and communication from one signal box to another was by simple electrical semaphore displays.

The system was gradually refined over the centuries from lessons learned from Courts of Enquiry after each railway accident. The accumulation of this hard-won knowledge made railway operators loath to change things and it was not until the 1960's that attempts were made to use transistor-based electronics to replace the aging hardware.

The design of highway traffic lights was influenced by the techniques used in railway signaling. Electromechanical relays were employed, allowing for little or no optimisation of timings.

Cars used thermionic valves for radios. Otherwise, their systems were mechanical or electromechanical. For good or ill, there were no speed cameras, parking meters, or a proliferation of CCTV's.

There were no commercial jet airliners until the deHavilland Comet went into service in 1952. Significant progress in complex aerospace systems had to await the transistor.

The electronics used in ships, aircraft and radar, that evolved out of wartime necessity, used thermionic valves. The equipment was expensive, heavy, bulky and unreliable compared with today's sophisticated systems, made possible by the transistor.

Medical

Many years ago, an aunt gave me an electric comb that had been in her possession for decades. It comprised teeth connected to an AA size 1.5-volt cell. The makers claimed that regular use of the comb would 'do wondrous things to your hair'. A label warned the owner to use only the maker's batteries!

Such quackery became widespread with the use of electricity. It traded on ignorance, awe and the fear of the new phenomenon. Perhaps this was similar to some of the therapies and medicines on which we spend so much money today or the behaviour of opportunists towards vulnerable owners of computers.

The American public seemed to be especially predisposed to quackery. A typical claim by one manufacturer reads:

> "There need not be a sick person in America (save from accidents), if our appliances become a part of the wardrobe of every lady and gentleman, as also of infants and children."

In the 1880's, Dr. Scott, an Englishman in America, was a notable advertiser and maker of "Electric Hair Brushes" and related shams. Initially these and other devices used slightly magnetized iron rods in their handles for their 'curative' powers. The connotations 'electric' and 'magnetic' were used to boost sales.[44]

For example, Dr Scott's original patent of 1st March 1881 contained the words:

> "The object of the invention is to secure within the interior of the brush one or more natural or artificial magnets, which, according to the belief of many persons, founded upon a theory of magneto-therapeutics which has become widely prevalent, have the effect of rendering brushes to which they are applied advantageous in use for relieving headache, preventing baldness, and other similar purposes". [sic].

(You may now take a breath)

It was not long before the list grew. It became common practice to describe electrical contraptions as cure-alls. Hair restoration, headache relief, laxative action, malarial lameness, rheumatism, diseases of the blood, and paralysis were just a few claims. Of course, each disease that was added to the claims opened up a wider market and increased the chances of a miraculous cure by at least one of them.

A further aid to sales volume was a note on Dr. Scott's hairbrush box.

> "In no case should more than one person use the brush. If always used by the same person it retains its full curative power".

(Impressive marketing)

Electrical appliances in medicine flourished over the following years with many instances of blatant deception. Vendors applied for patents so that they could use the term 'patented' to give credence to their dubious products. But there was also a growing and genuine interest in the potential benefits of electricity and

---

44 Perhaps similar to coffee venders today who apparently encourage the pronounciation Lar-tay' as in 'father' because it attracts greater sales than the correctly pronounced Lat-ae, as in 'trap'.

experimentation was often based on ignorance. In the UK and elsewhere, progress was hampered by a communication gap between the medical and electrical professions. Nevertheless, X-Ray and ECG machines, and other developments, began to bring great advances in medicine, although they were still relatively primitive until the transistor came along.

The first true hearing aids were ear horns. They were no more than a large funnel to focus sounds into the ear canal. Miller Reece Hutchison made an electric hearing aid in 1898 but we had to wait another twenty-two years for an aid with electronic amplification to become available. It was valve-based and called the Vactuphone. However, the size, weight and cost of the batteries, unsightly earpieces and carbon microphones held back the wide use of these products.

Transistor hearing aids began to appear in the USA in the 1950's

I remember attending a hearing aid symposium by the British Institution of Radio Engineers in the early 1950's. One of the hard-of-hearing delegates demonstrated his homemade, advanced hearing aid of which he was justly proud. He wore wartime metal headphones plugged into a large box of valve electronics and carried a case contained the batteries. Clearly, a transistor hearing aid of reasonable size and cost had not yet arrived in the UK.

Today, barely a day goes by before there is an announcement of an astonishing new advance in medical science. Most, if not all, of these advances have relied on the availability of powerful computers, which, in turn, relied on the availability of the transistor.

Military and Space

The defence industry made strenuous efforts to meet the requirements for weapon system electronics. Faced with valves and traditional wiring techniques it struggled to meet the challenges of extreme vibration, high acceleration, rough handling and wide temperature and humidity ranges. Also, missile and aircraft systems were in pressing need for small and lightweight equipment.

It is true that miniature, wired-in (rather than plugged-in) valves were developed that were less than two centimetres in length. These with other miniature components were crammed into minimum space but the assembly of the products was intricate and costly. They were difficult to repair and suffered from the same problems of poor reliability and high power consumption that were characteristics of the period.

Military equipment used low-voltage batteries or generators but its valves required high voltages. The conversion to these higher voltages was achieved with rotary machines or electromechanical vibrators of doubtful reliability. The arrival of the transistor would make possible small, switch-mode, very reliable power supplies.[45]

The harsh conditions within aircraft, tanks and rockets, favoured the use of electromagnetic[46] and electromechanical components over the more delicate valves. The construction of the Norden Bomb sight is just one example. (Plate 17). Nevertheless, aircraft and tanks did use valve-based electronics, particularly for radio, radar and sonar. Usually, short operational life expectancy made this acceptable. Even some of the ingenious and brilliantly engineered V2 rockets used transponders and radio control but everything needed to work faultlessly for only a few minutes.

Prior to and throughout World War II, radar, sonar, radio and computing systems were developed in a constant technical leapfrog with equally ingenious enemies. Considering the primitive nature of the components, by present day standards, the achievements in electronics from 1930 to 1950 were remarkable. However, electronic engineering was in its infancy and the design of

---

45 Examples of switch-mode power supplies are the cheap, plug-in mains adaptors that are so widely used in the home today.

46 Meaning iron cored components. Not to be confused with electromagnetic *waves*.

equipment in the UK was generally the province of the scientist. Their laudable achievements in conceiving war winning systems and devices are legendary but their paraphernalia was often badly designed and engineered compared with that in the USA and Germany, where the role of engineering was better understood. Thus, British equipment was ingenious but relatively unreliable.

A notable success however was the No.10, valve-based, microwave radio communication system, produced just in time for D Day[47] to provide communication across the channel with Normandy.

No. 10 Set aerial 1947

But, even after 1946, the lesson appears not have been learned, as illustrated with the design of Blue Danube.

Blue Danube was the UK's first operational nuclear weapon. It was a free-fall fission bomb with a yield of between 10 and 15 kilotons of TNT and carried by the aircraft of the V force. Production was limited to about twenty and they stayed in service until 1961. (See Plate 18). But its electronic fuse system contained a number of design errors that could have been avoided had it been properly engineered. Nevertheless, the electronics in the weapon was a good example of the resourceful use of miniature valves and components.

47 For younger readers, D Day was the day of allied military landings that took place along the coast of Normandy on June 6th 1944.

The V2 Rocket

Over 3,000 V2s were launched in anger during World War II of which some 20% were beam-guided. They travelled above the stratosphere, reaching a height of 80 km (50 miles). It is reported that Hitler commented on 22 September 1943, "It is a great load off our minds that we have dispensed with the radio guiding-beam; now no opening remains for the British to interfere technically with the missile in flight" [48]

From Glass to Silicon

This picture below shows the dramatic evolution of the diode power rectifier.

Rectification (the conversion of alternating current (AC) to direct current (DC) used valves or large selenium or copper oxide devices until the advent of the semiconductor rectifier.

The progress is even more remarkable because the working part, of the silicon rectifier, the silicon chip, is only a tiny fraction of the size of its casing.

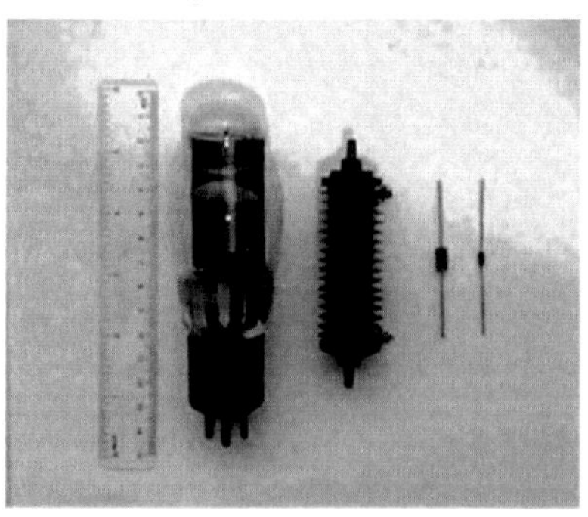

Left: A valve rectifier. circa 1930/40.
Centre: A selenium rectifier. circa1940.
Right: Two silicon rectifiers. circa 2000.

48 Irving, David (1964). The Mare's Nest. London: William Kimber and Co..

The Engineer

The role of an engineer was as misunderstood in 1940's as it is today. This is well illustrated by A.P.Rowe's[1] description in his fascinating book 'One Story of Radar'. To quote:

> "(At Malvern College) we sadly missed our model shop .... In the Autumn of 1942 formal approval was obtained for the building of what became known as our Engineering Unit..... By May 1943 a million bricks had been laid and a floor area of 72,000 sq.ft. was available.
>
> Engineering services comprising electricity supply, town gas and steam, compressed air, heating and ventilation, telephones and public address equipment were provided and all was made ready for the installation of machine shop equipment.............
>
> On 1 August 1943 workshop personnel moved into the Engineering Unit which began to provide an indispensable service to TRE scientists .....".

It is noticeable in the book that only scientists appear to have been invited to his 'Sunday Soviets' (weekly open discussion and brain storming sessions). There is no mention of engineers being present.

Engineers in the RAF were regarded as second class citizens;[49] a portrayal that the UK Government, civil service, finance sectors, careers advisors and particularly the media find it hard to come to terms with even today

Wikipedia again:

> "Many semi-skilled trades and engineering technicians in the UK have, in the past, called themselves engineers. In modern times this is seen as a misuse of the title, giving a false image of the profession".

Despite the notice seen on a faulty weighing machine in 2015 (see Plate 19), there are many signs of change for the better.

---

49 As described by F.J.Atkins in his book 'From the Ground Up'. Airlife Publishing Ltd. 1983. ISBN 0 906 393 21 3

Servo Mechanisms

Transformer principles were used to design synchros and servomechanisms for remote indication and control, instead of the use of valves. Being mechanical, they were able to provide greater power and could withstand the rigors of the military environment.

Typically, servomechanisms were used to control gun movements and aircraft control surfaces. They also rotated aerials and the electron beams in cathode ray tubes thereby improving the reliability of aircraft - mounted dish aerials and PPI systems (such as H2S).

Although still a valuable asset in the electronics repertoire of components, they are both costly and weighty and have been supplanted by semiconductor devices in many applications.

Chapter 3. The State of Play by 1950

1950 was a year of political and social tension. Great Britain was still struggling to recover from the deprivations and the crippling cost of the war. It was a year when petrol rationing ceased but food rationing was to last for another four years. There was a Cold War standoff between the West and East and the topic of the day was the nuclear deterrent and nuclear shelters. It was a year too when a war with Korea followed the *second* 'war to end all wars' Senator Joseph MacCarthy began his Communist witch-hunt and President Truman ordered the construction of a hydrogen bomb.

But there were also signs of optimism and modern living. The first successful kidney transplant took place in the USA and the first credit card in widespread use by the public, the Diner's Card, was distributed to just 200 people at the beginning of the year, growing to 20,000 by the end. Above all, 1950 was a turning point in electronics and marked the beginning of an age of solid-state electronics leading to the supercomputer.

We have seen the birth of electronics some 50 years previously and the accelerating progress that followed, limited only by the valve and its associated components.

In 1950, the transistor had been invented a little over two years before but it was not until the 1950's that commercial versions began to make an impact on the world. At first, they were low power, signal devices. Then the rate of advance continued exponentially, prompting the prediction made by Moores Law[50].

---

50 Moore's Law is variously described as:
'The number of transistors in an affordable CPU (or a dense integrated circuit) would double approximately every two years'. Or, 'The processing power of computers doubles every two years'. According to the major chip manufacturers, by 2021 transistors will shrink to a point at which will be no longer be economical viable to make them smaller. (James Titcomb's report. Daily Telegraph Business 26th July 2016).

Meanwhile, although thermionic valves had reached an advanced state of development, they still suffered from being large, hot, inefficient and unreliable. They plugged into valve holders so that they could to be replaced easily, thereby creating a prime source of trouble due to corrosion of the contacts. Equipment using valves needed high voltages. Its inefficiency generated heat and it was heavy and inherently costly. There was therefore a pressing need for a more efficient device to replace the valve.

We have seen that a typical black and white television set used some 20 valves and some of the valves had two devices in one glass envelope, a hint of the future Integrated Circuit. There was no colour television for the public until the transistor became available and hand-held remote controllers were yet to appear.

All radios were dependant on valves so that portable sets were large, heavy and greedy on batteries. Low power valves were available but most portable radios still required big and heavy batteries.

Computer memory was limited to the Williams Tube[51], invented in 1946, and the Delay Line memory in 1949. The Williams Tube used an electrostatic cathode ray tube[52] for digital storage and by 1948 it was able to store 1024 bits of information. Long-term data storage was restricted mainly to punched cards and paper tape until the Hard Disk Drives (HDDs) became available in 1956 and floppy discs in the late 1960's.

GPS navigation for the public, made possible by the invention of the transistor, had to wait for the turn of the century.

Railway signally and train control were carried out either mechanically or electromechanically. Electromechanical relays

---

51 The invention of Professor Fredrick C. Williams and his colleagues at Manchester University.

52 A cathode ray tube in which the beam was deflected by voltages on internal plates rather than by current through surrounding coils.

were used extensively for switching, particularly in telephone exchanges.

Apart from thermionic valves in radios, cars used mechanical and electromechanical components and required frequent attention. Figure 5 lists most of the features of typical modern automobiles. It brings home the incredible progress in car design. Those items that have been made possible, or significantly improved because of the invention of the transistor, are in bold type.

The first satellite, the Russian Sputnik, had to wait until October 1957 and the first satellite broadcast until 1962. Arguably, neither would have been feasible without the transistor.

The US Military used valve-based mobile phones during WW II but they were more like Dom Joly's contraption in 'Trigger Happy TV' rather than those of today. Little progress had been made by 1950 until the transistor made its impact. There was no cell-phone system.

Medical instrumentation was costly and large and, compared with today, limited in its operation. Its reliability, and that of domestic appliances generally, was poor.

Breakdowns and the expectation of failure of electrical equipment was a way of life. The transistor paved the way for extremely reliable components and better procedures for design and production so that failure has now become a rare occurrence; a fact that we take for granted.

There are outstanding examples of this incredible progress in reliability. Rosetta's Philae landed on a comet in November 2014. In July 2015, New Horizon probe, sent to the dwarf planet Pluto, coped with the vibrations of rocket lift off, temperatures as low as –225 degrees C, travelled for over nine years to its venue 3.67 billion miles from the sun, and arrived on time in working order. Then, in October 2016, Juno went into orbit round Jupiter after a five year journey.

Figure 5. The Influence of the Transistor on the Motor Car.

12,000 mile maintenance periods
ABS
Air Bags
Anti roll suspension
Automatic chokes
Automatic switching of lights
Automatic transmission
Break down triangles
Car radios
Catalytic converters
Central locking
Collapsible steering columns
Cruise control
Crumple Zones
Disk brakes
'Door Open' and seat belt warnings
Easyexit
Electric starters (replacing the front-mounted hand cranks)
Electrical windscreen wiper timers
Electric sun roof
Electric window control
Electric Windscreen washers
Electronic Code System
Electronic gear changing
Electronic ignition
Supplementary Restraint System
ESP (Electronic Stability Control)
Failed bulb warnings
Folding wing mirrors
Fog lights
Front windscreen wipers
Glove compartments (and lockable).
Headlight dipswitches
'Headlights on' warning signals
Headlight washer/wipers
Hydraulic brakes
Hydraulic engine valves
Interior air conditioning
Interior carpets
Interior Motion Sensor
Interior heating
Intruder alarms/engine disablement systems
Internally mounted batteries
Internal rear view mirrors
Intruder alarms/engine disarming
Locking petrol caps
Maximum speed control
Metallic paint
Moulded plastic trim.
Multi grade and detergent oils
Plastic body parts
Power steering
Proximity sensors
Quartz electric clocks
Radial tyres
Rain sensors
Rear seat armrests
Rear view mirror dippers
Rear window wipers
Rev counters
Road temperature indicators
Retractable and covert aerials
Safety glass windscreens
Sealed, no maintenance batteries
Seat adjustment
Seat belts
Side impact bars
Smart keys
Smart suspension
Speed limit control
Stoplights
Suspension
Synchromesh gears
Trip odometers
Tubeless tyres
Turn indicator flashers
Unleaded petrol
Visors
Visor-mounted vanity mirrors
Windscreen washers and heaters

It is easy to forget the inefficient state of a typical office in 1950.

The typewriter was the most important tool; a machine that had barely changed since the 1920's. Its QWERTY keyboard, that we still use, had been designed deliberately to slow things down in order to prevent the mechanism jamming. Typing was almost exclusively the job of a woman. Men were either incapable of typing or were not prepared to type themselves.

Modest typing mistakes could be overtyped using correction fluid or tape but significant changes to a letter or report required a complete retype. Often a lengthy document would need to be retyped several times.

Businesses used shared typists or maintained a 'first come, first served' 'typing pool'. When a single (not her marital status) secretary or typist was shared among several staff, she may set her own priorities, often to the frustration of those out of favour.

Secretaries took down the spoken word in shorthand and then typed out the content verbatim. Otherwise, a dictation machine would be used. Wax cylinder recorders for that purpose were still being used in many offices.

A small number of copies of a document (say up to four) was typed by inserting carbon paper between the blank sheets of paper but the readability of the copies deteriorated the further down they were in the layer. Multiple copies were obtained using a hand-operated stencil duplicator such as a Cyclostyle machine invented in 1890. The stencils had to be 'cut' by a typewriter, which called for yet another retype of a document. Up to 2,000 copies, of questionable quality, and prone to fading with time, could be printed using these machines.

Communication was by the telephone, the postal service, telex or fax. Telephones still used rotary dialing; introduced some fifty years earlier. Cynics claimed that the time it took to dial 999 to call the fire service, allowed the building to be well alight!

Administration departments relied on a large number of filing cabinets and card index systems for their data storage. Desktop mechanical calculators were used in accounts departments and technical staff used log tables, linear or spiral slide rules and Brunsviga machines. A replica of Barnes Wallace's office, on display at the RAF Museum in Colindale, is a good example of an engineer's office in the 1940s and demonstrates that things had hardly changed over the intervening years. (See Plate 20).

Drawings and diagrams were produced in large open plan areas that housed specialist draughtsmen and draughtswomen. Their work would be delivered as cyanotype blueprints or diazo prints[53]. This resulted in frustration and delays similar to those using the typing pool. The advent of the transistor was to make a gradual but dramatic change to the office workplace.

The Met. Office Super Computer.

In the 1950s, the Met. Office acquired an electrical desk calculator - which was cutting-edge technology at the time.

Today it uses a £100 million supercomputer.

53 Known as 'Whiteprints' - blue lines on white paper background. They faded if exposed to light for weeks or months

Chapter 4. The Solid State Revolution

Our story began with waves and has spanned little more than 80 years of amazing progress in what we now call electronics. From hereon it is difficult to keep pace with the progress of technology. It seems that one is overtaken by events even before the ink dries on the paper. New, more advanced, products and discoveries are announced and existing ones become history. What follows therefore can only reflect the state of things at the moment of writing. This is not, it appears, a new experience. At the turn of the nineteenth century the Czar of Russia, Nicholas II wrote,

> "Today these were the last word in science, tomorrow they were obsolete and had to be replaced"[54]

We have seen that some of the properties of semiconductors were observed throughout the mid 19th and first decades of the 20th century. They formed the basis of the photoelectric cell and the cat's whisker detector and were called metalloids. Their electrical conductivity (the ease with which they pass current) lies between a conductor, such as copper, and an insulator, such as glass. We have noted that the electrical resistance of a semiconductor varies with light intensity and can be made to emit light when electrically stimulated. Also its electrical resistance generally *decreases* with increasing temperature, which is opposite to that of a metal.

We recall that adding impurity atoms to a semi-conducting material is called 'doping'. It enables designers to change the electronic behaviour of the material to produce rectifiers, amplifiers, switches, sensors and energy converters that form the foundation of 'solid state electronics'.

Let us now look at the semi-conducting 'transistor' in more detail

---

54 From his manifesto calling for the limitation of armaments. Taken from "The Proud Tower" by Barbara W. Tuchman. Hamish Hamilton 1996.

The Transistor

The influence of the transistor is like a tiny seed that grew into a giant sequoia tree in half a decade. From the crude experimental model shown on page 7, the transistor has come to transform the way we live our lives. Without the transistor we would not have the Hydron Collider, huge supercomputers and phenomenal advances in our understanding of the human body and the universe.

So, how did it all begin?

Karl Braun

German scientist Karl Ferdinand Braun patented the first crystal rectifier in 1899 and in 1894 Indian scientist Jagadish Chandra Bose was the first to use a crystal for detecting radio waves. As we have seen, users of crystal sets manipulated a wire 'whisker' to find a semi-conducting spot on the surface of a crystal that was doped by nature. A copper oxide layer that formed on the surface of the wire probably helped this procedure although they would not have been aware of this at the time. Copper oxide and selenium rectifiers were used for power applications in the 1930s and germanium point contact and silicon crystal diodes for military radar and microwave relays in the 1940s.

Chandra Bose

After high purity semiconductor materials became available in the 1950s, inexpensive germanium *Junction* diodes, replaced *Point Contact* diodes as signal detectors.

In spite of early progress with the semiconductor *diode*, it could not amplify a signal. Many efforts were made to develop an amplifying semiconductor *triode* but these were unsuccessful because of limited theoretical understanding of solid-state materials.

So, although the vacuum tube triode valve followed rapidly after the invention of the vacuum tube diode, it took a further half a century to make a solid-state replacement for the triode valve. This huge technological achievement required an understanding of semiconductor action, based on a theory of solid-state, and progress in quantum physics.

The Very First Transistor?

From 1922, several people contributed to the evolution of a semiconductor 'triode'. Oleg Losev actually developed two-terminal, negative resistance amplifiers for radio in 1922 but, sadly, he died of starvation in the Siege of Leningrad in 1942, aged 38.

One can conjecture how history would have changed had he lived. Others from Austro-Hungary, America, France and Germany contributed to the evolutionary story but it was not until November and December 1947 that success was achieved in the USA.

John Bardeen, Walter Brattain and William Shockley were physicists at the AT & T Bell Labs in the United States. In 1947, they observed that when *two* gold point contacts were applied to a crystal of germanium, a signal was produced with the output power greater than the input. So, the solid state 'triode' had arrived and was called a 'transistor'.[55] The Bell Labs. team had benefited from the current knowledge of quantum physics to explain the properties of a semiconductor and the movement of electrons and holes in a crystal lattice.

In 1956 the three members of the team were awarded the Nobel Prize in recognition of their extraordinary work.

Hard on the heels of this momentous discovery was a growing understanding of semiconductor materials and their manufacture.

---

55 Lee DeForest had claimed that his invention, the vacuum valve, would never be replaced.

This made possible a rapid increase in the radio frequency at which transistors could work, the power they could handle and the mass production of complex silicon chips. Microprocessors and large capacity solid-state memory devices, the essentials for modern computers, were on the horizon.

Although it was a team of three that won Nobel Prizes it is Shockley who, by being the quickest off the mark in making their findings public, became the most prominent in the history books.

Like all historical references to the departed, recollections differ. Our fantasy bubbles about the past have a habit of being pricked by historians, each with their own interpretations. Impressions of Shockley are no exception although one thing is clear, he was a controversial character.

Time Magazine describes Shockley as "One of the century's most important scientists". But sadly his self-important, abrasive personality and unpopular views earned him few friends. Like many other inventors, he had to endure the wealth and power that resulted from his invention going to others. He died in disgrace, and, except for his loyal wife, Emmy, quite alone.[56]

In the face of diverse opinions and with very few exceptions, I prefer to keep a generous view of great men and women. There are many folk whose reputations remain unsullied because they achieve very little in life.

Courses on the subject of the transistor became popular in 1952. Engineers had been accustomed to designing their circuits with *voltage*-driven thermionic valves. They now had to get to grips with the transistor, a new concept of *current*-drive. As an engineer, I attended one such course at Borough Polytechnic given by a physicist. He started by writing Schrödingers Equation on the blackboard; "to calculate the wave function for an object", he said.

---

56 http://www.pbs.org/transistor/album1/shockley/ makes interesting reading about him.

Dutifully, we wrote everything down but I, together with the other attendees, were none the wiser by the end of the course.

This was a time when only a few understood or had even heard of this rarefied subject. I came away feeing rather like the fellow in the cartoon below.

Irrationally, I was not very keen on Schrödinger because he cost me a wasted course. Then recently I read the following by Chad Orzel[57], which did not improve my regard for the chap.

> "Schrödinger was almost as notorious for his womanizing [SIC] as for his contribution to physics. He came up with the equation that bears his name while on a ski holiday with one of his many girlfriends and fathered daughters with three different women, none of them his wife (who incidentally, knew about his affairs). His unconventional personal life cost him a position at Oxford after he left Germany in 1933 but he carried on living more or less openly with two women (one the wife of a colleague) for many years."

57 In his book, "How to Teach Quantum Physics to Your Dog." ISBN 978-1-85168-779-4

Schrödinger's Equation for Physicists

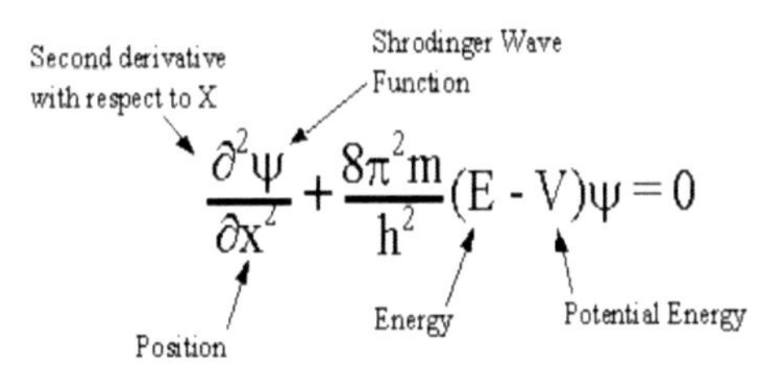

At the beginning of the twentieth century, experimental evidence suggested that atomic particles were also wave-like in nature. For example, electrons were found to give diffraction patterns when passed through a double slit in a similar way to light waves. Therefore, it was reasonable to assume that a wave equation could explain the behaviour of atomic particles.

Schrödinger was the first person to write down such a wave equation. Much discussion then took place on what the equation meant.

The eigenvalues of the wave equation were shown to be equal to the energy levels of the quantum mechanical system and the best test of the equation was when it was used to solve for the energy levels of the Hydrogen atom, and the energy levels were found to be in accord with Rydberg's Law.

It was initially much less obvious what the wave function of the equation was. After much debate, the wave function is now accepted to be a probability distribution. The Schrödinger equation is used to find the allowed energy levels of quantum mechanical systems (such as atoms, or transistors). The associated wave function gives the probability of finding the particle at a certain position.'[58]

---

58 Thanks to Ian Taylor, Ph.D., Theoretical Physics (Cambridge), PhD (Durham), UK

Perhaps after reading the above, readers may have some sympathy with my experience at Borough Polytechnic.

It was a considerable relief to find one of the Philips Company's pragmatic technical monographs that compared the valve with the transistor. From then on, all became clear.

The Valve and the Transistor (See Figure 6)

We have seen that the three electrodes of valves were called the anode, the cathode and the grid. The corresponding electrodes for the transistor are the collector, the emitter and the base.

Figure 6 shows two types of transistor, the NPN and the PNP. They both do the same thing but the NPN needs to be operated with its collector at a positive voltage with respect to its emitter, while the PNP needs to be operated with its collector at a negative voltage with respect to its emitter.

The first transistors were all PNP but both versions later became used in the design of circuits.

Figure 6. Circuit Comparisons of the Valve and the Transistor.

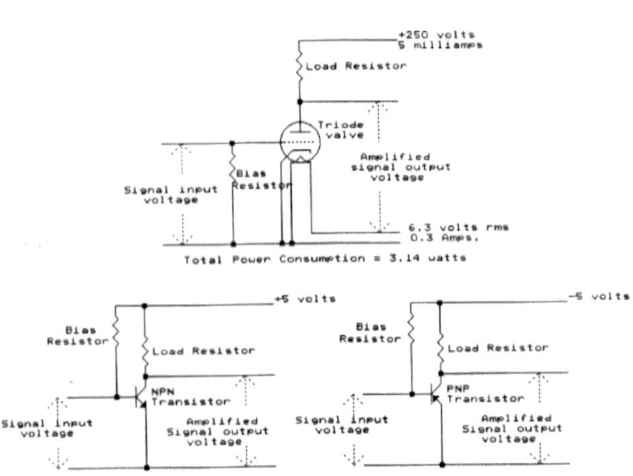

An Early Attempt to Make a Car Radio Using Transistors and a Printed Circuit Board.

Early in 1954, during my time with Murphy Radio in Welwyn Garden City, I was given a few of the first Point Contact transistors for experimentation and the ambitious task of building an in-car superheterodyne radio receiver. I was told to use no valves and to mount the components on a printed circuit board, at that time, a new and promising method of assembly.[59] My experience, described below, illustrates the pioneering nature of those early days.

To make the printed circuit board it was necessary to take a copper-clad board of reinforced phenolic resin and cover it with a thin layer of photo resist that was sensitive to ultra violet light. This was achieved by placing a drop of the resist on the centre of the copper and spreading it evenly over the surface by spinning the board on a 78-RPM gramophone turntable. The resist was then allowed to dry.

The circuit-wiring arrangement was drawn, or laid out with black tape, on a white background and a negative transparency of the drawing was made The transparency was then laid on the resist-covered board and 'exposed' by using ultraviolet light. The ultraviolet caused the resist to harden, according to the circuit pattern, making it resistant to the corrosive action of ferric chloride.

Now, the unhardened, unwanted area of copper had to be removed but because of the noxious fumes caused by the reaction of the ferric chloride with the copper, I had to

59 Instead connecting components with lengths of copper wire, the connection wires are replaced by tracks of thin copper strip on a copper-glad phenolic resin or glass-fibre board. The components, whose wires pass through holes in the board, are soldered to the copper tracks on the other side. The printed circuit method of assembly lends itself to mass production resulting in lower cost and far greater reliability. Complex circuits today, such as those used in computers, use multilayer boards.

carry the board to an outhouse. There, the surface of the board was gently rubbed with cotton wool soaked with the ferric chloride until only the circuit tracks remained.

Unfortunately it was found that the resist was susceptible to the effects of humidity so, on a damp day, it softened on its journey to the outhouse and I had to start all over again.

The outhouse was unventilated so the inhalation of the fumes during the ferric chloride etching process was most unpleasant. However, the excitement of the project made Logie Baird's sodium cyanide seem relatively harmless. But I am surprised that I still have a healthy pair of lungs.

The radio was assembled by drilling small holes to allow the wires of the components to pass though the board and solder them to the copper circuit tracks.

Because of the low cut-off frequency of those early transistors, it was only possible to receive a signal in the long-wave band. Nevertheless, many happy hours were spent with a colleague driving around the countryside in the firm's large Humber saloon to test the design. (There were plenty of volunteers to assist, particularly since the best reception seemed to be inside public houses!)

Later, germanium junction transistors replaced point contact transistors. They had a higher cut off frequency, which made it possible to cover the medium and short wave frequencies.

With this tedious process, it is likely that I can claim to be the first to person in the UK to make a transistor car radio.

The first commercial power transistors became available at the end of the 1950's from Pye Ltd., GEC and Texas Instruments. We experimented with these in an attempt to replace the unreliable mechanical vibrators in Blue Danube. But it was early days and I fear we destroyed lots of transistors in the process due the little understood 'runaway effect'[60].

Trying to replace valves with the available semi-conductors was always a challenge. Their white noise and temperature drift characteristics were often a limitation, particularly with those made with germanium. Because Silicon[61] semiconductors require the growth of very pure silicon crystals, it took until the mid-1950s for them to dominate the market.

We have seen that the size of the case of a semiconductor component belies the minute size of the active part inside. This fact led to the possibility of integrated circuits whereby many transistors are deposited on a single silicon chip.

The number of transistors on an integrated circuit chip grew at an astonishing rate. At first, a series of **T**ransistor **T**ransistor **L**ogic (TTL) and **C**omplementary **M**etal-**O**xide **S**emiconductor (CMOS) logic modules replaced modules like those shown on page 88 (Fig.9). They each contained tens of transistors. Then In 1965, as Gordon E. Moore had predicted[62], the number of transistors on a small thin wafer of silicon, less than 20mm (0.79inches) square, grew from 2,300 in 1971 to 2,640,000,000 in 2010. Then, in 2016 it was announced that all the books ever written in the world could now be contained on a chip the size of a postage stamp.

Imagine a system with 2,640,000,000 valves! Based on the assumption of 3 watts per valve, the valve equivalent of a silicon chip in 2010 would consume nearly 8,000 Megawatts; the same as

---

60 The path to a reliable power transistor was a tortuous one. It was found that the early products had a nasty habit of destroying themselves inexplicably. It took some time to discover that, if their collector to emitter voltage limits were even slightly exceeded (by a voltage spike on the mains electricity supply for example), thermal run-away occurred and destroyed the device. The effect was particularly troublesome in the design of power supply inverters. Eventually, ways were found to overcome the problem.

61 Not to be confused with silicone.

62 See footnote on page 62

the power consumption of 44 Large Hydron Colliders or the electrical power used by over 47 million people in the USA.

There was one serious drawback however. Solid-state devices like transistors were susceptible to the effects of strong electromagnetic pulses of radiation. The problem is illustrated dramatically in Jim Baggott's book, 'Atomic[63]'.

> "When Viktor Berlenko, a Soviet pilot with the 5 13th Fighter Regiment, defected with his MiG 25 Foxbat on 6 September 1976, the aircraft was examined in detail by the Foreign Technology Division of the US Air Force. The investigators were astonished to discover that it was full of obsolete valve technology. This simply didn't fit the natural presumption that the Soviets were far ahead in terms of technology development.
>
> The answer was quickly forthcoming. In an exchange of battlefield nuclear weapons, the electromagnetic pulse resulting from a nuclear explosion would completely disable aircraft filled with more sophisticated transistor technology. Indeed, the Soviets were ahead of the game. Better double the research budget."

But we are racing ahead of ourselves again.

(Pause for a coffee?)

Sound

The Bell Lab's invention in 1947 was just a demonstration. It was for industry to turn the prototype transistor into a reliable, mass-produced, low cost product so that from the mid 1950's transistors progressively replaced valves in manufactured products.

---

63 'Atomic. The First War of Physics'. Icon Books Ltd. 2009. ISBN 978-184831-044-5

In 1952 hearing aids became hybrid[64] in which manufacturers replaced the valve with a transistor in just the output stage. Then, in January 1952, Microtone's first all-transistor hearing aid appeared. Those of Maico, Unex and Radioear soon followed. Some 225,000 hearing aids were sold in the United States In 1953. Of these, 100,000 were all-transistor, 75,000 were hybrid, and 50,000 used only valves. Then, just one year later, the market sold 335,000 hearing aids of which 325,000 were all-transistor models.[65] They were very much smaller, unobtrusive and used much less battery power.

From the 1970s microprocessors, provided advanced features for hearing aids and offered greater amplification. Products progressively reduced in size with advanced, inconspicuous in-the-ear aids.

Portable radios however were the leading products to use transistors. It is said that they became the most popular electronic communication device in history with billions manufactured during the 1960s and 1970s.[66] Intermetall, demonstrated an early model at the Dusseldorf Radio Fair in Germany in 1953. Apparently it performed better than valve radios but it is not clear in what respect. It was not produced in any quantity.

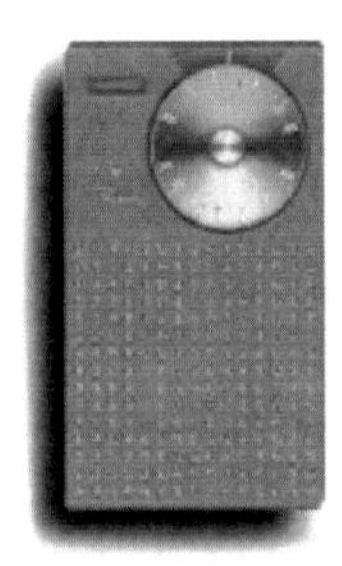

Regency TR-1

Texas Instruments and Industrial Development Engineering Associates jointly created the first commercially available transistor radio in November 1954. Called the Regency TR-1, it contained 4 NPN transistors and one diode and covered the frequency range 640 to 1240 kHz; part of the Medium Waveband. Despite its limitations, about 150,000 TR-1s were sold.

64 A combination of valves and transistors

65 According to the American Hearing Aid Association

66 Quentin R. Skrabec, Jr. McDonald's (2012). The 100 Most Significant Events in American Business: An Encyclopedia. ABC-CLIO. p. 197. ISBN 0313398631.

Raytheon
8-TP-1

Then, in 1955, Raytheon made a big step forward in sound quality with its 8-TP-1 radio. It used 8 transistors but still provided reception only in the Medium Wave band.

Sony TR-55

August 1955 saw a small Japanese company, Tokyo Tsushin Kogyo,[67] export a 5-transistor radio to the USA under the brand name 'Sony'. It was called the TR-55,. It was the first company to build a radio using its own make of transistors and other miniature components. The success of the product caused Tokyo Tsushin Kogyo to change its name to SONY. The TR-55, and subsequent models, resulted in over 6 million Japanese transistor radios being sold in the United States by 1959.

Sound Recording

A Cassette Tape

The reel-to reel home tape recorder of the 1950s 60's and 70's (see Plate 21) was replaced by the Compact Tape Cassette, introduced by Philips in 1963. It also served for data storage for the first microcomputers. From the early 1970s the cassette, together with the Long Playing vinyl disk (LP) were the two most common formats for prerecorded music.

The small size of the cassette, together with transistor-based electronics, made recorders so small that they could be carried

67 The Tokyo Telecommunications Engineering Company

around in a pocket. Like the transistor radios, they popularised the use of headphones and were described as 'personal'. In particular, Sony introduced a series of portable audio cassette players called the 'Walkman' in the late 1970s and these florished during the next three decades.

From the year 2000 the cassette was overtaken by the compact disk (CD), the minidisk and later by the USB drive, mobile phone, Internet downloads and streaming.

Figure 7 demonstrates the dramatic increase in the storage capacity of each system, the last six of which depend on the transistor.

Figure 7. Audio Recorders

	Size	**Playing Time**
Reel-to-reel tape	128 mm diameter x 10 mm thick	25 minutes
Audio tape cassette	100 mm x 65 x 8 mm thick	45 minutes per side
CD	120 mm diameter x 1 mm thick	74 minutes.
DVD	120 mm diameter x 1 mm thick	120 minutes
Minidisk	68 × 72 × 5 mm	80 minutes
USB stick	40 x 12 x 4 mm	1500 hours
Modest Mobile Phone	115 x 60 x 10 mm	>360 hours

Television

The first tentative use of transistors in television sets appeared in the late 1950's. They used a combination of valves, discrete transistors and semi-conductor diodes. Valves had to be included because the semi-conductors at the time could not cope with the high voltages needed for the line scanning circuits. Also the existing small signal transistors were not able operate at VHF frequencies.

Sony TV8-301

The Sony Corporation is recorded as the first company to introduce an all-transistor television. Released in 1960, it was called the TV8-301 and had a five inch black and white screen. Nine new transistor designs solved the problem of high frequency tuning. Although expensive, the product was premature and unreliable.

Subsequently, several manufacturers competed in the market for transistorised television receivers so that by 1972, a typical all-transistor, colour television receiver contained 80 transistors, 40 solid-state diodes and 2 integrated circuits (which themselves contained several more transistors). Later, solid-state circuits replaced even the capacitors and inductors.

Video Recording

Modern video recording systems demonstrate the astonishing progress brought about by the transistor.

The first video recorder, called VERA, was developed experimentally by the BBC from 1952. But it was not completed until 1958 by which time the Ampex VRX-1000 (later renamed the Mark IV) had arrived and further work on VERA ceased.

The BBC design recorded on 20½-inch diameter reels of magnetic tape traveling at 200 inches per second. VERA used a stationary head so that the magnetic impressions of the video signal were created along the length of the tape in a similar way to audio tape recording. The Ampex machine, on the other hand, used a rotating head, a considerable step forward. This formed a diagonal striped magnetic pattern across the tape as it moved forward. The technique was adopted for all subsequent video tape recorders.

Using transistors, Sony introduced the first machine for the home in 1964 followed by Ampex and RCA one year later. Then, in the 1980's, competition occurred between three incompatible formats

- Sony's Betamax, first introduced in 1975, JVC with VHS in 1976 and Philips V2000 in 1978. Eventually, a marketing triumph for VHS made the Sony and Philips products obsolete. The view at the time however was that Betamax was a technically superior system.

Solid-state electronics popularised the home videocassette recorder but with products having incompatible recording formats. They all recorded in colour and included their own tuners and clock timers. The machines were about the size of a briefcase, cost around £300 at the time and their tape cassettes could retain up to four hours of recordings. Later, the use of DVD's instead of tape provided only two hours of recordings but had the advantage that they were randomly accessible and did not have to be wound back.

Tape cassette player/recorders have now been replaced in the home by Digital Video Recorder/Players and, at the time of writing, a typical DVR costs under £200. It can have three internal tuners and can record up to four programs simultaneously. Using its internal hard disk, it can store over 600 hours of recordings and, like the DVD machines it requires no rewinding.

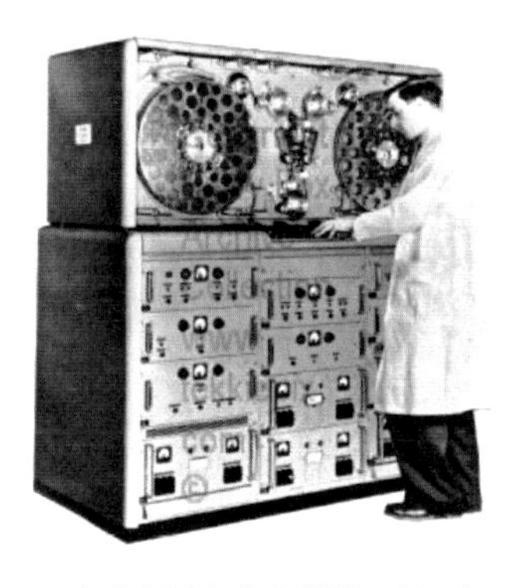

The BBC Black & White 'Vera' Video Recorder. 1952.
½ inch tape at 200-inches/sec. on 20½-inch diameter reels.
Playing time: Approx.15 minutes, using 15000 feet of tape.

The Ampex 1000a Black & White Video Tape Recorder (later called the Mk IV). C1958.
(As displayed in the London Science Museum).
2 inch tape at 15 inches/sec. on 10-inch diameter reels.
Playing time: 1 hour using 4,500 feet of tape. Weight: 665 kg.

The Philips N1500 Videocassette recorder. 1972.
Size: Approx. 430 x 300 x 82 mm (17 x 12 x 3¼ inches)

The Humax FVP-4000T Digital Video Player/Recorder. 2015
Size: 280 x 200 x 48 mm (11 x 8 2 inches)

Medical

Transistors began to make their mark in medical equipment in the 1960's. Two early examples, a 1975 defibrillator and a kidney dialysis machine are shown in Plates 22 and 23.

The incredible pace of progress has provided the medical profession with an enormous range of relatively low cost, small and powerful electronic devices. The modern intensive care units (ICU's) in hospitals resemble mission control centers as can be seen in the following pictures.

Intensive Care Monitoring

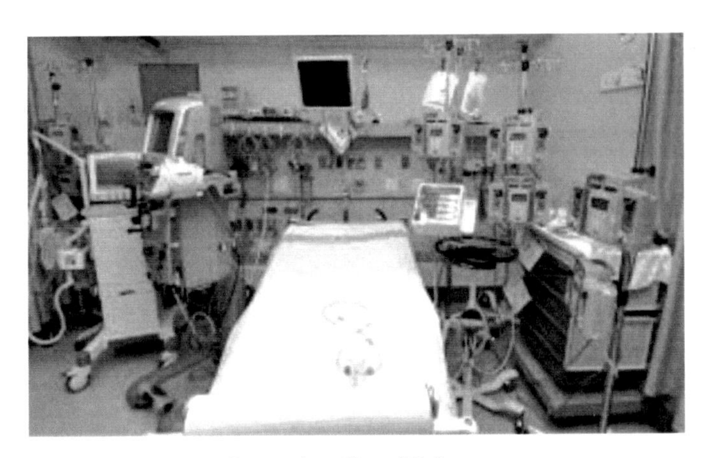

Intensive Care Unit

Today, solid-state electronics enables us to embed Implantable Cardioverter Defibrillators (ICDs), into the chest. They continually monitor the heart via electrodes and if they notice a dangerous heart rhythm they can deliver the following treatments:

- A series of low-voltage electrical impulses (paced beats) at a fast rate to try and correct the heart rhythm. (Pacing).
- One or more small electric shocks to try and restore the heart to a normal rhythm.
- One or more larger electric shocks to try and restore the heart to a normal rhythm. (Defibrillation).

They can communicate in real time with a medical center and manufacturers, such as Medtronic and St. Jude Medical make readings available to doctors via company-run gateways.

Probably the most far-reaching developments, brought about by the solid-state computer, are the compilation of the human genome, stem cell therapy, artificial Intelligence (AI) and robotics.

Today, robotic surgery (even by remote control via the Internet) is proving safer than by the surgeon's hand with less blood loss and shorter recovery times.

The Transistor's Progenies

The birth of the transistor brought about an industry that generated many discrete semi-conductor devices.

Silicon was known to be a better material than germanium for the future development of semiconductor devices, one of its advantages being that it suffered less from the effects of temperature. However, it took many years to find a way to purify crystals of Silicon in sufficient quantity to make a transistor. It was Gordon Teal of Bell Labs who achieved this in April 1954, prompting Texas Instruments to start production only four weeks later. Silicon then paved the way forward for the incredibly complex semiconductors of today.

As early as 1906, H J Round had discovered the principle of the light emitting diode (LED) using silicon carbide crystals. But it was the stimulus of the transistor and the development of silicon semi-conductors, that gave us modern LED lighting diodes, based on Gallium; another semi-conductor.

Appendix 3 lists some of the many other discrete semi-conductor products that the transistor spawned - its prodigies.

Soon transistors were combined to produce series of analogue and digital modules. At first, these were just transistor circuits encapsulated in resin but an integrated circuit comprising several transistors on a single wafer of silicon soon replaced them. (See Figure 9).

These integrated circuits were used primarily for digital logic circuits and were the elements of early computers. Analogue modules, such as the ubiquitous integrated 741 analogue amplifiers, also became widely used.

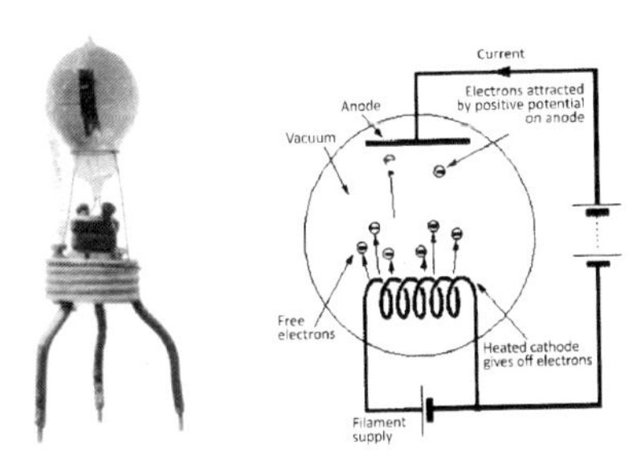

1. Ambrose Fleming's thermionic diode valve

2. A very early Marconi transmitter at the 2MT station in Writtle, Essex.

3. The 2LO transmitter

4. The 2LO transmitter

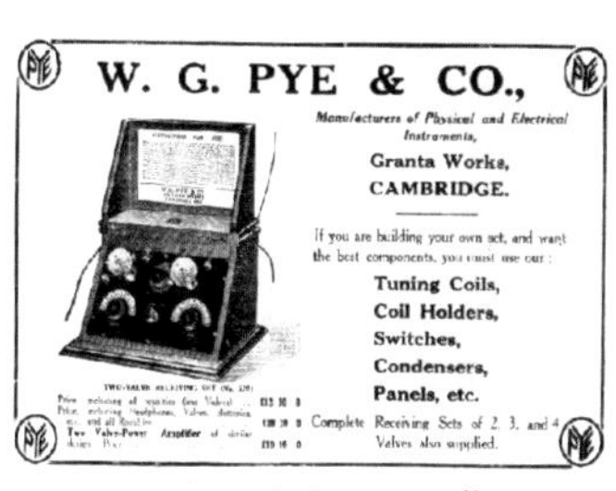

5. An early battery radio

6. The Philco PT-44, mid 1930's 'All America Five' 5-Valve superhet.

7. The Marconi-Stille razor wire tape recorder.

8. The first known photograph of a moving image produced by Baird's 30-line 'Televisor', circa 1926. (The subject is Baird's business partner, Oliver Hutchinson).

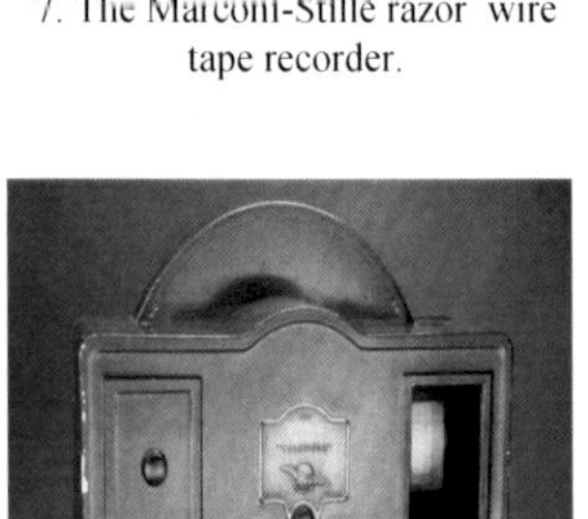

9. The 30-line 'Televisor'. The tiny screen is on the right.

10. The Brunsviga stepped gear calculator, 1930s/1940s.

11. The author's 9" tube, 405-line television in 1947. Note the lethal 5,000 volt mains transformer and capacitor in on the right.

12. The author's completed DIY television. 1947

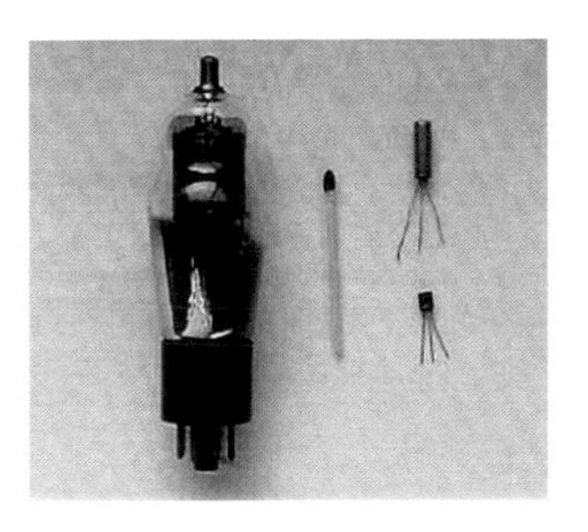

13. A triode valve and a match, compared with an earlier and later transistor. The active areas of the transistors are minute compared with the size of their cases.

14. A 1950's valve ampilifier with its integrated circuit equivalent and an G.P. 10 pence coin.

16. **ENIAC**, the first electronic general-purpose computer. It was Turing-complete, digital, and capable of being reprogrammed to solve "a large class of numerical problems." Wikipedia

15. Babbage's Difference Engine. (At London Science Museum)

17. The Norden Bomb Sight (Shown at the Computer History Museum in Mountain View, California.

18. Blue Danube, the first nuclear weapon issued to the RAF, The impressive casing obscures the unreliability of the valve-based electronics inside its nose.

19. 'An engineer has been called'

20. A replica of Barnes Wallace's Office c 1940

21. The Philips 4,500 Reel-to-reel 1970 tape recorder with four tracks.
It had three tape speeds , 7.½, 3¾and 1⅞ inches per second.
Measuring 49x34x18 cm, it weighed 9.5kg and used 26 transistors.

22. 1975 External Defibrillator.
As displayed in the London

23. 1968 Cardiff Kidney Dialysis Machine.
As displayed in the London Science Museum.

24. The Fujitsu K Supercomput

Custom Chips

These are special-purpose silicon chips containing large numbers of transistors designed to serve particular applications. Considering the list in Figure 8, it might have been easier to name applications in which transistors did *not* play a part.

Clocks and watches are particularly impressive examples of the influence of the custom silicon chip. In the mid 19th century, the best clocks gave an error of 0.1 of a second per day. Then, just prior to the advent of the silicon chip, this improved to 0.001 of a second per day. Now, using quartz crystals and solid-state electronics, watches with errors of 0.0001 of a second per day cost only a few pounds and their prices have less to do with the cost of their mechanism than the quality of the casework and the brand name. The world's most accurate clock has an error rate of less than one second in 15 billion years.[68]

As the cost of the microprocessor and the microcomputer reduced it was often easier to use *them,* instead of custom chips and program them for each task. (Not always trouble free however).

Figure 8. Custom Chip Applications

Air conditioning	Elevators	Pacemakers (Internal)
Bank service tills	Flat screens	Pedometers
('Holes in the Wall')	Greenhouse control	Photocopiers
Blood pressure meters	Hearing aids	Printers (Laser & Inkjet)
Boiler control	Human genome	Radio and television
Calculators	iPads	studios
Car components	LED and 'low energy'	Radio receivers
Cash registers	lighting	RFID and bar code
CAT Scans	Mains adaptors	systems
Clocks & watches	Medical Devices	Security systems
Computer games	MRI	Smart phones
Curtain pullers	Mobiles phones	Solar energy
Defibrillators (Internal)	Modern elevators	Space travel
Digital photography	Motorway signs	Telephones
Domestic appliances	Pocket radios and players	Television receivers
Electric toothbrushes	On-line banking	Test equipment

68 From the Daily Telegraph 22nd April 2015 (Nature Communications)

Figure 9. Early Logic Modules

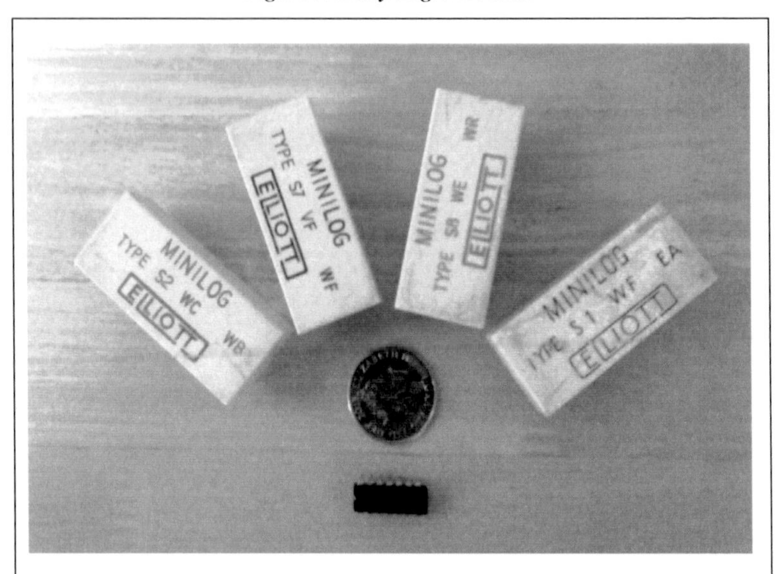

One of the early attempts to take advantage of the relatively small size and low power dissipation of the transistor was a range of digital logic modules such as the Philips 'Combi Elements". These were little more than a set of standard components and transistors mounted on a printed circuit board, molded in resin and enclosed in a box measuring 2"x 3/4" x 5/8". Because of their simplicity, unskilled outworkers could be used to assemble the modules.

The Elliott company produced a similar range of logic modules. Examples of these are shown above, compared in size with a ten pence coin. The integrated circuit below it replaced them all.

De Havilland Propellors Ltd. developed a hybrid computer called Anatrol in 1961[69] using analogue operational amplifiers and multipliers. These, together with capacitor storage were reconfigured every second, using a Post Office-type uniselector. In spite of its slow speed the product was adequate for oil pipeline operations.

Another approach, again by De Havilland, was based on the

69 'System Engineering in Theory and Practice'. M James and G S Evans. Journal of the Brit.I.R.E. January 1961.

'Tinkertoy' project. Here, modules comprised stacked wafers on which transistor chips were mounted. Resistors were scribed on the wafers using an electron beam machine. A Digital Equipment PDP 8 minicomputer was used to control the scribing process.

Television receivers contain custom chips and are now more like computers. They are described as 'smart'. This allows software updates to be received over the air that, in effect, modify the design of the set.

Computers

The astonishing progress in computing power that transistors have brought about is more like the stuff of science fiction. The smallest and fundamental element in this progress is the microprocessor.

The Microprocessor

A microprocessor is a Computer Processor (CPU) on a wafer (a microchip) of silicon. It can be regarded as the "engine" that drives all modern digital computers by obeying a set of programmed instructions.

An early CPU was built in 1955 using 200 individual transistors. Seventeen years later, in 1971, engineers Federico Faggin, Ted Hoff, and Stanley Mazor, working for an emerging three-year-old company called Intel, produced the first microprocessor CPU on a single silicon chip. It was called the 4004 and was used primarily to perform simple mathematical operations in a calculator. This revolutionary design

contained 2,300 transistors on a wafer area of 3 x 4 mm and sold for $60. Since then, microprocessors have become more powerful and faster as the number of semiconductors they contain has increased in line with Moore's Law. Yet, they are faster, small, and cheap.

By 2015, a microprocessor chip could have as many as 10,000,000,000 transistors[70] and single chips contained several CPU's. Some idea of the small size and the huge number of solid-state devices can be seen by viewing a mobile phone screen or television flat screen through a magnifying glass.

Microprocessors are now found in aerospace systems, avionics, meteorology nuclear physics, engineering, transport systems, telephones, tablets, consumer electronics and the home. Clusters of microprocessors form the 'brains' of data centers, super-computers, communications products, and other digital devices.

Progress has been such that the Intel i7-6700K microprocessor used in a good home computer contains 4 cores, over one billion transistors and has a speed of 4.2 GHz. At the time of writing it costs £300.

Microprocessors are relatively inexpensive and very reliable because there are few external electrical wire connections. Their function is to accept data, act upon it according to programmed instructions stored in memory, and then output the result. The simplified diagram overleaf shows the pride of place of a microprocessor in a typical home computer system.

The potential of the microprocessor was recognised by the UK government in 1978 when it introduced an imaginative grant-aided scheme called MAPCON.[71] Its purpose was to encourage the

---

70 Transistor counts are relevant to the subject of this book. However, they are not necessarily a reliable measure of the performance of a computer.

71 Microprocessor APplication CONsultancy.

British manufacturing industry to appreciate the benefits of micro*processors* for the control of its processes.[72]

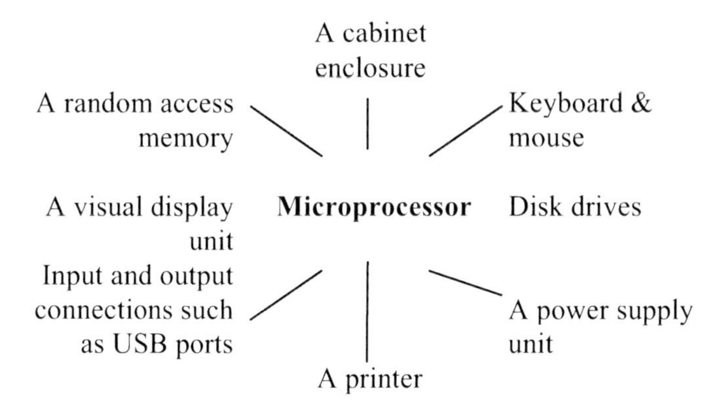

Some significant successes where achieved. For instance, my company saved one food manufacturer millions of pounds[73] because of a MAPCON-funded feasibility study costing less than 1% of that saving. However many managers found it hard to understand, and adjust to, an unfamiliar concept of software programming.

Later the scheme was extended to include micro*computers* in the office. But, as with the manufacturing processes, the take-up was tentative. For example, I advised one CEO to buy a microcomputer and have a screen and keyboard on his desk, only to be met with the response, "I'm not going to be a ruddy typist!" On the other hand, the chairman of another company that had just received a very large contract insisted on *immediately* replacing his company's well-tried manual accounting and stock control system with a computer. It was not a surprise that the company went into receivership soon after it was installed.

---

72 The use of the word microprocessor was misleading. It was intended to mean microcontrollers that were based on microprocessors.

73 In today's currency values.

The Microcontroller

A typical *microcontroller* is a rudimentary, low cost but powerful computer on a small printed circuit board. It uses a *microprocessor* chip, that may contain over 400,000 transistors, serving as the CPU, memory and input/output controls.

Although microcontrollers are genuine computers, they are as small as a credit card and can cost as little as £4. They can be programmed with little experience and make it easy for both novices and professionals to deploy. They connect easily to monitor screens; T/Vs, keyboards, mice and other devices making them useful even as a Personal Computer. For example, they are capable of browsing the Internet, word-processing, making spreadsheets and playing simple games. But in particular, they are a low cost device for controlling things in the outside world such as electronic music, weather stations, lighting, heating, security systems, domestic appliances, simple robots, motion detectors, motors and actuators.

As with the early days of the personal computer, several companies have entered the field with their own versions of a microcontroller. Time will tell which will survive to dominate the market but, at the time of writing, the most popular products are the Raspberry Pi, the Arduino, the Galileo and, the BBC micro:bit.

The Raspberry Pi 3

The Raspberry Pi series of credit card-sized, single-board computers, is a British development by the Raspberry Pi Foundation.

The foundation's original intention was to promote the teaching of basic computer science in schools. However, its products are now widely used with some six million having been sold in the three years to 2015.

The Raspberry Pi 3 is a powerful computer with a quad core 1.2GHz processor and 1 gigabyte of memory. It uses less than 13 watts of power and is only 85 x 56 x 17mm in size.

The Raspberry Pi 3

The Arduino.

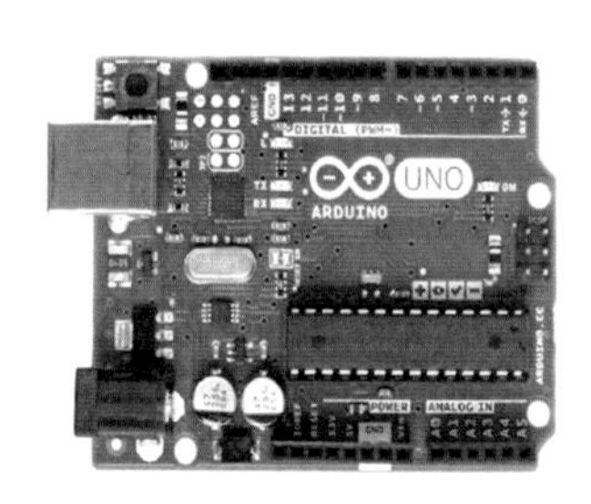

The Arduino
(Also called the 'Genuino' outside the USA).

Arduino is an *open-source*[74] computer hardware and soft-ware company that designs and manufactures a range of 'Arduino' microcontrollers. The first Arduino was introduced in 2005, as a low cost, simple tool for novices and professionals Like the Raspberry Pi, it can interact with the outside world of sensors and actuators such as simple robots, thermostats, motion detectors, lights, and motors.

74 Meaning, "The universal access via a free licence to a product's design or blueprint, and universal redistribution of that design or blueprint, including subsequent improvements to it by anyone". (Wikipedia.)

The Intel Galileo

The Intel Galileo is an Arduino-certified, open-source hardware and software development system. It is a development board that complements and extends the Arduino series giving it greater computing power. The Galileo is designed to be compatible with a wide range of Arduino Uno R3 products.

The Intel Galileo

BBC micro:bit

This is a credit card-sized (4 cm by 5 cm) successor to the 'BBC Micro' of 1981. Although more limited than the microcontrollers described above, it is designed to complement them. A micro:bit has been freely distributed to every year-seven pupil in the UK.

It is interesting to compare, below, the BBC micro:bit with the BBC Micro that was launched thirty five years earlier.

	The BBC Micro of 1981	**The BBC micro:bit of 2015**
Processor Type	8 bit	ARM Cortex M0 32 bit
Process Speed	2 MHz	16 MHz or 32.768 MHz
Storage	Cassette tape or 5½" floppy disk.	Via USB

Operating System	BBC BASIC	N/A
Display	PAL or NTSC television receiver.	5x5 LED matrix with 25 red LEDs
Connections	Printer 1 MHz bus, second processor interface.	Bluetooth LE, microUSB
Networking	None	2.4 GHz Bluetooth low energy
Power	50 Watts mains	2 x AAA batteries
Size	409 mm x 358 mm x 78 mm	52 mm x 42 mm x 11.7 mm
Weight	3700 g	8 grams
Price	£335 each	(1 Million free)

The Microcomputer

A microcomputer is a small machine, designed for use by one person at a time, in which the central processing unit (CPU) is contained in one or more silicon chips. Commonly called a personal computer (PC) or (confusingly) 'micro', it takes one of two forms, the 'Desktop' (which usually stands *under* a desk) and a portable, battery/mains-powered 'Laptop'.

From the 1970s, the microcomputer made a dramatic change to world of computing. Using advanced microchip technology, these low-cost machines evolved to use microprocessors containing over a billion transistors and largely replaced minicomputers.

Small, cheap and reliable, the microcomputer now has the computing power of the earlier computers that filled an entire room. They have huge large internal and external storage capacities and connect online, form networks and use various input and output devices, such as laser or inkjet printers, CD-ROMs, DVDs and USB flash drives.

History often leaves us in doubt about who was the first in anything. Each nation has its own opinion and the microcomputer is another example of the 'winner' being the one to attract the most public attention.

As with Colossus, secrecy and confidentially sometimes prevent early disclosure of an invention. It is likely therefore that the first operational microprocessor was designed and developed from 1968-1970 to operate the movement of the control surfaces and to display pilot information for the US Navy's F14A 'Tom Cat' fighter jet. However, Ed Roberts became recognised as the 'father of the microcomputer' when, in 1975, he designed the Altair 8800, using an Intel 8080 microprocessor. Its success was guaranteed when William (Bill) Gates and Paul Allen supplied the BASIC Interpreter software for the machine.

Several makes of Microcomputers that could to do useful work such as accounting, database management, and word processing, were soon to follow. By 1973, they used an 8-bit processor such as the Intel 8080 or Zilog Z80 and an S100 bus. Also, at the end of 1974 Intel's rival, Motorola, introduced its M6800 microcomputer system, with an initial clock speed of 1 MHz

In 1983, another company called Apple Computers produced Lisa, the first microcomputer to utilize a Graphical User Interface (GUI)[75]. It also used a mouse. Lisa had a clock speed of 5 MHz, 1 Mbyte of RAM, a hard drive, two floppy disk drives and a 12-inch black and white monitor and cost $23,900 in 2016 money values.[76]

---

75 The system used to enable users to communicate visually and easily with the computer. Windows is an example.

76 £16,900 at the time.

Lisa was not a great commercial success but it paved the way to the modern laptop and desktop computers.

A scene-changing event took place in 1981 when IBM introduced its Model 5150, a well-engineered and modular designed microcomputer. It was the first to be described as a 'Personal Computer' or 'PC'. The great name of IBM and the advent of Lotus 1-2-3 software made it attractive to business users. By 1984, it was sold in the millions.

Throughout the 1980s and early 1990s the three applications, 'Lotus 1-2-3' (for spreadsheets) 'dBase' (for databases) and 'WordPerfect' (for word processing), became the forerunners of today's 'Office' suites. Then, with Microsoft's release of Windows 3.0, in May 1990, and the falling prices of machines, the microcomputer was here to stay, both in business and in the home.

Lotus 1-2-3 comprised a spreadsheet with an elementary database facility and graphical charts. It quickly ousted its predecessors, VisiCalc, Multiplan and Supercalc. The Lotus 1-2-3 spreadsheet

More About Early Micros.

Other contenders may have a claim to be the first to build a microcomputer. For instance, in 1962, Lincoln Labs. produced the first desktop microcomputer with a keyboard and screen, called the 'Linc'. It was designed for use by biomedical technicians in 1962. About 2,000 were made and they sold for $40,000 each.

The Guidance Computer (AGC) in Apollo 11 Command Module of 1969 enabled astronauts to enter simple commands by typing in pairs of nouns and verbs. AGC had approximately 64 Kbytes of memory and operated at 0.043MHz, making it more basic than the electronics in a modern washing machine.

As we have seen, Intel's groundbreaking 4004 microprocessor chip, introduced in 1971, grouped all the parts of a computer on one silicon chip and was able to read and respond to instructions.

The French filed a patent in 1973 for a solid-state microcomputer using a microprocessor and from the 1970's, several companies built and sold microcomputers. Here are a few that some readers may remember with affection or otherwise.

Sinclair ZX81, MITS (Altair), BBC (Micro), Cromenco (System Three), SCELBI-8H, Commodore (PET), Radio Shack, TRS-80, Apple (Atom), Atari, Tandem, North Star (Horizon), Altos, Rair (black box), Amstrad, IMSAI, Southwest Technical Products Corp, Morrow Designs and Ohio Scientific, Ernst Steiner (es65), Intertec (Superbrain), Epson (QX-10 & PX-8). Of these, the BBC Micro is regarded as the first home, personal computer in the UK.

Early Operating Systems

From 1973, the operating systems adopted by most manufacturers were either CP/M, (for a single machine) or MP/M (for a network). A number of companies produced their own simple application programmes to run on their operating systems, now (regrettably) called 'apps'.

Many suppliers emerged with their own brand of personal computers, based on component modules that were copies of those of the IBM PC. The BIOS[77] of the IBM PC was reverse engineered[78] and most models used DOS and, later, Windows for their operating systems. These products became known as 'compatibles' or 'clones'. Arguably there was little to choose between them. As usual, their survival depended on the strength of publicity, their reliability and the quality of after-sales support.

Manufacturers that did not follow the IBM microcomputer architecture, fell by the wayside. Apple was an exception however. Its computers had their exclusive design and operating system.

77 The BIOS or Basic Input/Output System is a program permanently stored in a Read Only Memory (ROM) microchip or microprocessor. Its purpose is to automatically start and prepare the computer system for use after it is turned it on; a process called as 'booting'.

78 Analysed and copied.

Reminiscent of the early days of wireless and television, the modularity of the IBM PC encouraged enthusiasts to build their own machines. Many of them chose to use the free, open source[79] Linux as their operating system instead of buying Windows.

The first PC compatible mouse, based on the earlier tracker ball, was introduced in 1982.

By 2015, PC's were widely available with clock speeds of 4 GHz Random Access Memories (RAM) of 8 Gbytes, 250 Gbyte solid-state drives (SSDs) and hard discs (HDD) of 2 terabytes. Their price of under £400 was a staggering drop from the £23,900 cost of the relatively primitive Lisa in 1983.

As the cost of microcomputers decreased and more of them were used in the home, a new generation of computer-numerate users emerged. At the same time, they were vulnerable to hard selling and dubious extended warranties and support services. At first, the machines were used mainly for utilitarian purposes such as letter writing and home accounting. But as a new generation of young and old (Silver Surfers) users emerged, emailing, texting, surfing the Internet, manipulating photographs, playing games and using the social media predominated.

The transistor made the Cellnet system possible. Mobile phones proliferated, and as they have increased in complexity they are described as 'Smart'. By 2010, they and the 'Tablet Computer' vied to satisfy the mobile computing market. Products developed rapidly and now, only a few months elapse before new, more advanced models appear, rendering earlier models obsolescent

Smart phones and tablet computers now feature touch-screen displays, cameras, microphones, accelerometers, an on-screen, pop-up display of a qwerty keyboard and batteries that last for

---

79 Open source software is freely available to anyone.

hours or days between recharge; all in a flat enclosure less than one centimeter thick.

Some products comprise a tablet that is detachable from a physical keyboard allowing them to be used either as a conventional laptop or as a tablet.

At the present time, the mobile market is split between two main product design philosophies. One uses Android (a Google, Linux-based, operating system) that is intended mainly for smart phones and tablet computers. Brands that use Android include Samsung Nexus, Acer, Asus, Toshiba and Sony. The other uses Apple's OSI Operating System for its Apple iPad and iPod Touch.

By 2012 Apple had sold 100 million iPads but by 2014 the iPad's market share was 36% compared with 62% for Android tablets. In 2015, the worldwide market share for Android smart phones was 82.8% and for iOS types, 13.9%.[80]

'Smart' is a word now widely used by the market to give sales appeal to almost anything. The makers of mobile phones, watches, television receivers and domestic appliances use the word 'Smart' to cover anything from access to the Internet (for the Internet of Things) to the connection of one device to others. For example, the South Korean company, LG, says of its smart television, "(It) detects your external devices such as Blu-ray™ [SIC] Player, Home Cinema System, Soundbar, Games Consoles, and USB sticks, and displays them on your LG Smart with a simple pop-up".

More About The Tablet

Typical tablet computers have from 16 to 128 Gbytes of flash memory and an SD Card reader. They do not have any other external connections for data storage or sophisticated peripherals. Their processing speeds are comparable with laptops.

80 http://www.idc.com/prodserv/smartphone-os-market-share.jsp

The Flat Screen

By the end of the 1980s Cathode Ray Tube (CRT) screens that could clearly display 1024 x 768 pixels[81] in colour were widely available and affordable.

The first 18inch standalone (and expensive) LCD flat screen (the Eizo L66 Flexscan) appeared in the mid-1990s.

By the turn of the century, with falling prices, the flat screen competed with CRT monitors so that waste-recycling centres became inundated with discarded CRTs (as well as floppy discs).

The Mini Computer

Minicomputers have processing powers, storage capabilities and physical sizes that are smaller than those of a mainframe but larger than those of a microcomputer. They were first sold to small and mid-size businesses for general business applications, to large enterprises for department-level operations and to academic institutions. Typical of the genre was the Elliott Company's 900 series of 1965. Desk-sized, and floor standing, it was large, by later developments. Elliott Automation used a 903 machine to control the first computer-based train describer. It was installed in the Leeds central signal 'box' and became the first use of a general-purpose digital computer involved in to the control of signalling in British railway.[82]

In 1965 the Digital Equipment Corporation (DEC) produced the most widely used minicomputer computer in the world at the time. It was the PDP-8 series. The first to arrive in the UK was used to

81 Dots on the screen
82 'A Computer Controlled Train Describer'. L.J Bental, G S Evans and A J Thomkins. The Radio and Electronic Engineer. Vol 38. No.6 December 1969.

control an electron beam welder for the manufacture of microcircuits.[83]

The PDP-8 was followed by the PDP-11 in 1970 and the VAX series in 1977. Their disk units comprised 14-inch platters, in a heavy 10.5" x 16" x 36" enclosure. They gave a storage capacity of only 120 Mbytes using fixed media Winchester disk drives, consumed 480 watts of power and weighed 102 kilograms. This contrasts with forty years later when a *many terabyte* 3½-inch disk drive requires only 20 watts. The VAX series became known as superminis.

These early machines did not boot (start up) automatically in the way in which we are now familiar. The operator had the tedious task of setting a sequence of numerous switches on a front panel every time the computer was switched on; a kind of human ROM.[84] Also, to communicate with the computer, a Teletype machine was used. Its keyboard or paper tape reader provided a means of data input and its printer or paper tape punch for data output.

The central processing units (CPUs) of these early minicomputers were built using integrated circuits and discrete semiconductors. But rapid progress in silicon chip fabrication and printed circuit techniques resulted in single chip processors and double-sided (later multi-layered) printed circuit main boards.[85] It became possible therefore to build smaller and more powerful *microcomputer* PCs using highly integrated microprocessor chips and fewer associated components.

The increased computing power of microcomputers in the 1970's made the minicomputer obsolescent although in recent years they

83 'Computer Control of the Manufacture of Electron Beam Processed Microelectronic Modules', G S Evans & Matcher-Lees. The Radio and Electronic Engineer 1967.

84 Read Only Memory

85 Known as Mother Boards.

have re-emerged as mid-range servers, such as the IBM AS/400e, and can form part of a network.

The PDP-8 Minicomputer

A DEC PDP-8 resembling a Sci-fi 'Brain' (1966)

The PDP-8 was the first computer to be mass-produced. It sold as a system for a price of around $18,000.[86] Its 12 bit single accumulator could address up to 32KByte 12 bit words. With just eight basic instructions the PDP-8/E could execute simple ones in 1.2 microseconds and complex memory referenced instructions in 4 microseconds. This gave the machine a rating of about a 0.5 million instruction per second (MIPS). The standard model had 4,000 words of Random Access Memory and a Teletype Model 33 ASR was used for data input and output.

86 The exchange rate in 1962 was 2.8 US dollars to the GBP, so $18,000 was £120,000 at 2013 pound values.

The Elliott Automation 900 series of Minicomputers

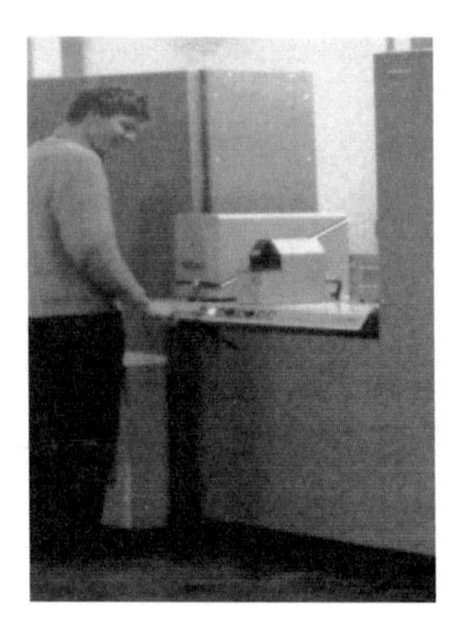

The Elliott 903 Floor Standing Minicomputer controlling a 'Train Describer' (1968).

The 903 and 905 computers of the 900 series (first introduced in 1963) were civilian versions of the military computers. The machines used transistors in plug-in packages.

The military 920B, was used in the Nimrod Mark I aircraft and the 920M in the RAF Jaguar and in army tanks. Typically they had an 18-bit word ferrite core store with a 6 microsecond cycle time although they *could* have an 18-bit word ferrite core memory of up to 126K. Usage was by paper tape for input and output and a Teletype .

The 903 version was a desk-sized machine. There was no disk for storage although magnetic tape was available. Peripherals could include a plotter, a line printer, magnetic tapes, industrial interfaces and displays for plant monitoring.

The computer proved to be popular with universities and colleges as a teaching machine, with small research laboratories as a scientific processor and as a machine for use in industrial process control. Some 1,000 machines were sold.

Programming languages that were available included Algol, Basic, Coral, Fortran II and the SIR assembler"

The Mainframe Computer

IBM. zEnterprise Mainframe Computer

Following the success of J. Lyons LEO-1, the 1950s saw the rise of digital computers in business. These computers came to be known as mainframes. They were large and required much space, air conditioning and a staff to run and maintain them. Because they were programmable they could be used for a wide variety of applications such as scheduling, data base management, communication and accounting.

Mainframe computers led to centralisation within a business and completely transformed the way things were done. They were, and still are, seen in cinema films as 'props', showing conspicuous revolving tapes to depict a dramatic scene in a computer room.

The number of manufacturers of mainframe computers grew over the next two decades, each competing in price and power. IBM led the field with its 700/7000 series and later the 360 series, trailed by Burroughs, Univac, NCR, Control Data Corporation (CDC), Honeywell, General Electric RCA and others. Then, as IBM dominated the scene, some fell by the wayside.

The popularity of the minicomputer and microcomputer networks meant that new mainframe installations were confined mainly to use by financial institutions and government administration. However in the late 1990s big business found new uses for their existing mainframes as they moved back towards more centralised computing. Also, the growth of the Internet and the advent of Big Data needed large mainframe systems and vast amounts of data storage.

By 2014, IBM was offering the zEnterprise EC12 mainframe that had a 5.5 GHz, six-core processor[87] and 3 terabytes of **R**andom **A**ccess **M**emory. (This contrasts with today's average personal computer that has some 4 to 8 Gigabytes of RAM.[88]) It involved 1,000,000,000,000 semiconductors.

The Supercomputer

IBM Blue Gene P

The Big Daddies of the computer world are the supercomputers. They are needed for 'number crunching' such as required in science, engineering and meteorology, where very fast processing speed and huge data storage are essential. They cost tens of millions of pounds sterling, are truly huge in size and represent the 'cutting edge' in computing. They can have up to half a million CPUs, which contain a staggeringly high number of transistors.

Supercomputers are the subject of worldwide competition to evolve the fastest machine as judged by 'Linpack'.[89] Japan led the race briefly in June 2011 when Fujitsu installed its K Computer at the RIKEN Advanced Institute for Computational Science campus in Kobe, Japan (Plate 24). It is capable of performing more than 10

87 It can have up to 101 cores

88 Our brains contain around 80 to 100 billion cells called neurons - the tiny switches that let you think and remember things. That is roughly equivalent to 10 to 12.5 Gigabytes. Yet, it weighs only about three pounds.

89 A benchmark application developed to solve a dense system of linear equations.

petaflop calculations per second, equivalent to one million personal computers (PCs) tied together.[90] The machine has 548,352 central processing units and comprises 672 to 800 cabinets full of circuit boards. It costs $10 million (£6.2 million) annually to run and is said to use the power of 10,000 homes.

The American Titan Supercomputer of 2012, made by Cray, has 299,008 central processing units, a RAM of some 700 terabytes and storage of 40 petabytes. Its speed is 17.59 petaflops. Its cost is 97 million dollars and it uses 8.2 Megawatts of power.

The British Meteorological Office uses a £97 million "Cray XC40" supercomputer. With 480,000 central processing units (120,000 times more RAM than a top-end smart phone) and a storage capacity of 17 petabytes, it can operate at 16,000 trillion calculations per second. It weighs in at 140 tonnes.

The Chinese, Sunway TaihuLight, has an awesome 1.31 petabytes of RAM, 10,649 processor cores, is capable of 93 petaflops.[91] and consumes 15 Megawatts of electrical power. At the time of writing, it is most powerful computer in the world. At a said cost of 273 million US dollars it represents China's 167th supercomputer compared with 165 in the USA and 12 in the UK.

Supercomputers are used for various heavy-duty processing jobs including climate and earthquake prediction, disaster prevention, cosmological calculations, medical research, stock market predictions, complex product modelling, nuclear research and weapons development.

Another kind of 'supercomputer' uses the Internet to combine the computing power of the millions of Personal Computers throughout the world. Each owner is asked to download a fraction of a project's data for analysis and send back the results. This work takes place while the computers are not otherwise unused. Cancer

---

90 Prof. Dongarra

91 93,000 trillion calculations per second. (Compared with 500,000 with a 1966 computer).

Research and the Search for Extra Terrestrial Intelligence (The SETI project) are just two of the areas that use this technique.

The Met. Office Supercomputer

A Met Office statement in 2014 read: "The new supercomputer, together with improved observations, science and modelling, will deliver better forecasts and advice to support UK business, the public and government. It will help to make the UK more resilient to high impact weather and other environmental risks. The first phase of the supercomputer will be operational in September 2015 and the system will reach full capacity in 2017".

Users of a Supercomputer.

Like a 'best kept secret' the proliferation of supercomputers is impressive. Here are just a few of the organisations that own or use one.

NEC, Fujitsu and Hitachi Sony, Nippon Denso (Electric Equipment), Toshiba (using a four-processor Cray Y-MP8), Mitsubishi Electric, NTT, Sharp. Nissan Motors, Toyota, Honda Research, Mazda Motor Corporation, Isuzu Motor, Daihatsu, Suzuki Motor Corporation, Mazda, Hino Motor, Toyo Rubber Tyre Company Sumitomo Rubber Industries, National Super Computer Center in Guangzhou China, DOE/SC/Oak Ridge National Laboratory United States, DOE/NNSA/LLNL United States, DOE/SC/Argonne National Laboratory United States, Swiss National Supercomputing Centre (CSCS) Switzerland, Texas Advanced Computing Centre/University of Texas United States, Forschungszentrum Juelich (FZJ) Germany, DOE/NNSA/LLNL United States, Leibmiz Rechenzentrum Germany, The National Supercomputing Centre, Wuxi, Nr. Shanghai.

As we end this chapter, it becomes clear that, without the invention of the transistor, this story would have stopped in the middle of the 20th century

Chapter 5. Our Future?

When I first set pen to paper (or rather fingers to keyboard) for this chapter I aimed to forecast the effect of the transistor on the future. However, with every passing day, the press contained news of advances in technology that made my words outdated. I decided therefore, to limit the chapter to a few topics that may have most influence on our lives.

Forecasting is a risky business and has a habit of rebounding on the prophet. Even Jules Vern and H.G.Wells had their moments. Winston Churchill said of Professor Lindemann (Lord Cherwell), "the Prof's brain is a beautiful piece of mechanism" and apparently the Professor agreed with him. However, the Prof predicted that the Germans could not possibly have a large rocket, saying, "to put a four-thousand horsepower turbine in a twenty-inch space is lunacy". Shortly after, the first one fell outside my previous house in Staveley Road, Chiswick. He also thought the idea of space travel was nonsense.

One reads[92] that a British Parliamentary Committee of 1878 decided that Thomas Edison's light bulb was not worthy of the attention of practical or scientific men. Then, in 1895 Lord Kelvin (William Thomson) maintained that "heavier than air flying machines are impossible" and the chairman of IBM in 1943 believed there was "a world market for maybe five computers". It seems that most of the incorrect past predictions were too pessimistic.

With the above in mind, it is reassuring that if some of *my* predictions do not happen I will be in illustrious company. So, here are some things that the transistor and the phenomenal progress in solid-state physics, will bring to us in the years to come.

---

92 The Giant Book of Bathroom Dumbology by Geoff Tibballs. Constable, London. 2014

Phones, TVs, Tablets, Cameras and Things

The efficacy-to-size ratio of the programmable microprocessor poses a severe problem to manufacturers who have to forecast their future Audio/Visual products.

All the devices in Figure 10 are capable of a wide range of features. The question is, which one will dominate the market and where does one risk their investment. That is the problem that faces companies like Apple, Microsoft and Samsung. I would hate to be in their shoes at this time.

Figure 10. Future Investment?

Feature	Phones	Cameras	PCs	Tablets	Televisions
Taking Single Shot Images	Yes	Yes	Yes	Yes	No
Video Recording	Yes	Yes	Yes	Yes	Yes
Video Playback	Yes	Yes	Yes	Yes	Yes
Sound Recording	Yes	Yes	Yes	Yes	Yes
Sound Playback	Yes	Yes	Yes	Yes	Yes
Internet Access	Yes	?	Yes	Yes	Yes
Phoning/Mailing/ texting	Yes	?	Yes	Yes	Yes
Radio	Yes	?	Yes	Yes	Yes

Telescopes

Telescopes of the future will enable us to find life on other planets in the universe. In 2015, 'Nature' claimed, "The building blocks of life have been found for the first time in a distant star system". Larger and more powerful terrestrial radio telescopes are appearing that comprise several small dishes or aerial arrays over great areas of land, linked together and coordinated using computers. These and space telescopes, like the Hubble, would be inconceivable without the transistor.

Quantum Computers

The next quantum leap (literally) will be progress in the field of Quantum Mechanics (QM), the Alice in Wonderland world of the atom. It was knowledge[93] of Quantum Mechanics[94] that gave us the diode, the transistor, the electron microscope and Magnetic Resonance Imaging (MRI) machines. Future understanding of the cosmos, on the vast scale and of the atom in the minute scale, will open up new challenges and will lead to a unified 'Theory of Everything'. A quantum computer could give us immense processing speeds and data security, with unbreakable ciphers. It will help us to understand the most complex scientific problems.

It is claimed that a fully programmable quantum computer that can solve simple algorithms has been made already[95]. No doubt, there are many 'back rooms' where progress is continuing at a rapid rate.

A special class of quantum mechanical applications is related to macroscopic quantum phenomena such as superfluid helium[96] and superconductors[97]. These have remarkable properties. In a superconductor, an electric current, once started, may flow forever. It is called a persistent current. "The persistent current was first identified by Onnes and attempts to set a minimum period for their duration have reached values of over 100,000 years."[98])

---

93 'Knowledge', not necessarily 'understanding'.
94 Or 'quantum theory'.
95 University of Maryland
96 Having zero viscosity.
97 Having zero electrical resistance.
98 Wikipedia

More on Quantum Mechanics

"Spooky, bizarre and mind-boggling are all understatements when it comes to quantum physics. Things in the subatomic world of quantum mechanics defy all logic of our macroscopic world. Particles can actually tunnel through walls, appear out of thin air and disappear, stay entangled and choose to behave like waves."

Swapnil Srivastava December 2012.

"Anyone who is not shocked by quantum theory has not understood it".

Niels Bohr. (Nobel Price in Physics in 1922).

"I think I can safely say that nobody understands quantum theory".

Richard Feynman. (Nobel Price in Physics in 1965).

So, according to Richard Feynman, I am on safe ground whatever I write here but it is not a subject for light bedtime reading.

Although we can describe the different ways that matter behaves when it is very, very small (quantum physics) and when it is large (classical physics), physicists are struggling to explain the transition from one to the other. However, let us shrink down with some of Alice's 'Drink Me' and see what Quantum Physics has to tell us.

1) Quantum simply means a quantity or an amount. In this context, it means the smallest amount of a physical quantity that can exist independently, especially a discrete quantity of electromagnetic radiation.

2) Quantum Mechanics deals with physical quantities and their interactions with energy at the atomic and subatomic levels.

3) Light behaves both like **particles** and like **waves**. This Quantum Duality is like the Holy Trinity; they mean the same thing but which of them is used depends on the context of the moment.

4) Photons are discrete units of electromagnetic radiation including light. (They have been likened to an aerosol paint spray where each droplet represents a photon. Coverage can be increased either by a concentrated burst or by several small accumulating bursts).

5) Qubits and Superposition. (The ability for a particle to be in two places at the same time).

We know that today's digital computer operates by using **'bits'**. That is, '0's and '1's. Bits can be represented by holes in a punched tape, places in a storage element in computer memory, a magnetic element on a hard disk or a 'burned' spot on a CD or DVD. Using '1's and '0's the processor's in our computers can only perform one calculation at a time.[99]

A quantum machine uses **'qubits'**. These are not only '0's and '1's but also all points in between and all at the same time. Qubits are represented by atoms, ions, photons or electrons and their respective control devices, all of which can work together to act as computer memory and processor. This is called a ***superposition*** of '0' and '1'. Using superposition, a quantum computer can perform many calculations at the same time.

Thus, a quantum computer has parallelism. It can have these multiple states simultaneously, making it millions of times more powerful than today's most powerful supercomputers. "This parallelism allows a quantum computer to work on a million computations at once, while your desktop PC works on only one".[100]

6) Heisenberg's Uncertainty Principle says, that it is fundamentally impossible to know the position and speed of a particle at the same moment. For example, if you press a key on your qwerty keyboard you can measure accurately the position of the key and the speed at which you pressed it. But In the quantum world, you cannot do both at the same time.

7) Determinism is Probabilistic.

Until a measurement is made, the particle/wave essentially exists in all positions! This paradox was put forward famously in the form of the Schrödinger's cat in the box 'thought experiment'.

8) Quantum Entanglement. (Strong correlation between particles that would be nonsensical in our everyday world).

---

99 That is because they operate serially; a step at a time. 'Parallel processing' is possible however by using a number of processors at once.

100 David Deutch

Quantum mechanics allows one to *think of* interactions between associated objects at a speed faster than that of light (the phenomenon known as quantum entanglement), of frictionless fluid flow in the form of superfluids with zero viscosity and current flow with zero resistance in superconductors. Quantum entanglement does not actually enable the transmission of normal information faster than the speed of light, yet the particles appear to do so because they are always connected and can behave as one.

With quantum entanglement, elementary particles behave in a bizarre manner. They become linked so that when something happens to one, the same thing happens to the other, no matter how far apart they are. This bizarre behaviour of particles in which they become inextricably linked together has been described as "spooky action at a distance."

Even though each particle has a lot of information about the other, they do not send messages back and forth. There are no messages between the particles saying, "I'm going down, therefore, you must go up" and waiting for the particle to receive the message.

9) Quantum Tunnelling.

Let us assume that a hurdle in an obstacle race requires a certain amount of energy from a hurdler to jump over it and to reach the other side. If the hurdler does not have that amount of energy, he or she will only hit the hurdle and bounce back, stop, or become ensnared in the hurdle. But if shrunk to quantum level, the hurdler would reach the other side of the hurdle even without the required amount of energy. He or she would *tunnel* to the other side.

10) Teleportation is making an object or person disintegrate in one place while a perfect replica appears somewhere else. At present, it is a science fiction occurrence popularised by the phrase, 'Beam Me Up Scottie'. But IBM warns that 'science fiction fans will be disappointed to learn that, for a variety of engineering reasons, no one expects to be able to teleport people or other macroscopic objects in the foreseeable future, even though *it* *would not violate any fundamental law to do so'* (writers italics). The last ten words are tantalizing.[101]

John Gibbin gives a fascinating insight into the feasibility of teleportation.[102]

A Further Word on the Subject.

"Quantum computing won't run *on* your phone - but maybe some quantum process of Google's will be key in training the phone to recognize your vocal quirks and make voice recognition better. Maybe it'll finally teach computers to recognise faces or luggage. Or maybe, like the integrated circuit before it, no one will figure out the best-use cases until they have hardware that works reliably.[103] It's a more modest way to look at this long-heralded thunderbolt of a technology. But this may be how the quantum era begins: not with a bang, but a glimmer".

Clive Thompson. Daily Telegraph Business 20th May 2014

Graphene

Another Alice in Wonderland phenomenon is Graphene. It is a type of carbon (an allotrope of carbon) but only one atom thick.

Graphene has many extraordinary properties and may eventually replace silicon for computers. Here are a number of uses for the material that have been suggested.

---

101 IBM Research 11/9/2014

102 John Gribbin 'Computing With Quantum Cats'. Transworld Publishers 2014.

103 The same applied to the introduction of the laser, which was described as a device looking for a purpose.

i) Your phone battery would need only five seconds to fully charge
(See; http://www.inquisitr.com/555843/graphene-batteries-offer-5-second-iphone-charging/)

ii) Simple and rapid purification of radiation contaminated water with easy disposal of the waste.
(See: http://news.rice.edu/2013/01/08/another-tiny-miracle-graphene-oxide-soaks-up-radioactive-waste-2/)

iii) A greatly improved tennis racquet.
(See, http://gizmodo.com/5975246/graphenes-newest-trick-is-improving-your-tennis-game).
No doubt other sports equipment will also benefit from the use of graphene.

iv) Extremely thin, unbreakable touch screens for phones, computers, television and other displays (New York Times).

v) Inexpensive, very high quality and almost weightless headphones
(See: http://gizmodo.com/5990520/scientists-have-made-graphene-earphones-and-they're-amazing).

vi) Greatly improved hearing aids

vii) Very high power supercapacitors to replace batteries.
(See: http://earthtechling.com/2013/02/graphene-supercapacitor-battery-thats-not-a-battery/).

viii) Sea water refining for human consumption.

ix) More efficient, longer lasting and cheaper electric light bulbs.
(See: http://www.bbc.co.uk/news/science-environment-32100071)

Perhaps the most exciting and promised applications for graphene are for the generation of electricity, both by solar and mechanical means. For example, wallpaper that converts light into electrical power, and roads and footpaths that derive power from human activity, that would otherwise be wasted.

But these are early days. Jesus de La Fuente, the Chief Executive Officer of Graphenea, writes,

> "In terms of how far along we are to understanding the true properties of graphene, this is just the tip of the iceberg. Before graphene is heavily integrated into the areas in which we believe

it will excel at, we need to spend a lot more time understanding just what makes it such an amazing material.

Unfortunately, while we have a lot of imagination in coming up with new ideas for potential applications and uses for graphene, it takes time to fully appreciate how and what graphene really is in order to develop these ideas into reality. This is not necessarily a bad thing, however, as it gives us opportunities to stumble over other previously under-researched or overlooked super-materials, such as the family of 2D crystalline structures that graphene has born"[104]

Medical

It is unlikely that the Human Genome that followed the discovery of DNA would have been available to us without the computer. With its use, gene therapy will play an increasing part in preventive and curative medicine. This, together with the use of stem cells, should lead to a reduced demand on the health service.

Hearing aid design has far to go. Laser-signalled hearing aids with a larger frequency range are already under investigation and induction charging rather than battery replacement will result in permanent in-ear hearing aids that never have to be removed. There will be better discrimination against background noise and damaged parts of the inner ear will be repaired using stem cells and solid-state devices.

Due to solid-state electronics, advances in the manufacture, monitoring, repair and endurance of the organs of the body (and the mind) have been made possible and will advance rapidly. Pacemakers, defibrillators and cochlea implants are just three indications. Solid-state processors and memory will be implanted into the body to cure malfunctions of, or to augment, the brain.

---

104 www.graphenea.com/pages/graphene-properties

With the use of Robotics, AI and the Internet, more surgical procedures will be carried out remotely and safely.

Medical treatment and medication will become largely self-diagnosed, administered and tailored to the individial.

It will become possible to communicate with animals' brains by way of solid-state sensors and implants.

3-D Printing

3-D printers use a string of heat-melted material instead of ink. Layers of the material are printed (deposited) progressively one on top of the other to build up a solid 'printout' (3-D object). Just as various coloured inks can be used with a conventional printer, so a 3-D Printer can deposit various materials such as plastic, metal or even food. A great advantage of fabricating an object with a 3-D printer is that complex spaces inside solid objects are easy to achieve.

Like conventional printers, 3-D printers are controlled with software programs.

3-D printers are in their infancy. Yet, they are used already for making body parts such as cartilage and bone replacements. Huge versions are building houses. 3-D printers are used in the space stations to make parts, so averting the problem of carrying a large number of different spare parts. They will make a huge impact on the home, the retail trade and the way we live.

The Internet and The World Wide Web

In 1964, I attended a conference at Edinburgh's Herriot Watt College, entitled, "The Impact on Computing of User's Needs on the Design of Data Processing Systems". I remember two things in particular about that conference.

First, at that the top of the delegate's 'wish list' was a low cost Random Access Memory. (The most advanced RAM at that time used slow and expensive magnetic cores). Then there was astonishment at witnessing a demonstration of a real-time data link with a university in the USA. We all waited with baited breath for the text message to arrive.

Just one year earlier J.C.R. Licklider, working for Bolt, Berenek and Newman, produced a memorandum proposing the concept of an 'Intergalactic Computer Network' that had several of the features of the modern day Internet. Later in that year he moved to the Defence Department's Advanced Research Projects Agency (ARPA) and convinced it of the importance of such a system. His vision resulted in the ARPANET for military use.

My company was given access to the Arpanet as part of a defence contract. One day I noticed several members of the staff standing around a computer screen in embarrassed silence. A programmer had stumbled by accident upon a password that caused masses of US Government top-secret information to roll down the screen. Everyone was frozen with a look of guilt and disbelief until the programmer had the presence of mind to quickly log off. My company may have been the first hacker, albeit a guiltless one.

We need to distinguish between the Internet and the World Wide Web. The Internet is a mesh[105] of connections between large numbers of 'hubs'. A message can be sent from one hub to another via any, or many, of the other hubs. This means that if one hub pathway is busy, the message is diverted via others. By splitting up the message into small sections, each section can find its quickest route to its destination. On arrival, the sections reunite to reconstruct the message. This speeds up the delivery enormously.

The World Wide Web (www.), on the other hand, is the software the runs on the Internet. It is a global information protocol for

---

105 Hence the word, InterNET

sending and receiving data via computers connected to the Internet, such as website pages or emails.

A rough analogy is that the Internet is like the hardware of physical computer and the Web is like the operating system software.

By the end of 1990, Sir Tim Berners Lee had developed the three main elements of software necessary for a working Web. These were the **H**yper**T**ext **T**ransfer **P**rotocol (HTTP),[106] the **H**yper**T**ext **M**arkup **L**anguage (HTML) that we use for designing web sites and a browser. We now have several makes of browser, the most common being Internet Explorer, Firefox, Yahoo, Google, Microsoft Edge and Bing.

The future of the World Wide Web is unclear. New ways to cope with a massive increase in traffic and the problem of its misuse will occupy future minds.

The Cloud

"The Cloud" is so called probably because it appears to be 'out there' and nebulous. It means that our PCs can access and make use of a huge network of commercially owned computer resources (or servers) with vast memories. So, it is really a use of the Internet and is not nebulous; it is tangible.

Cloud resources are managed and protected by several competing organisations; some are free and others are paid for according to the number of gigabytes of storage used. They claim to provide a completely safe and 'everlasting' depository for your data, accessible only by the use of your password. The recent abrupt closures of some cloud suppliers puts ‘everlasting’ in doubt however.

Google, Apple and Microsoft are just three examples of Cloud resources.

---

106 Which we often use as http:// followed by www. to access a web site.

Today, almost everyone knows about computer software. However, in order to appreciate how the cloud works, the following summary may be helpful.

Software comes in two flavours. One is the *data* that we generate, download or stream[107] such as letters, emails, accounts, photos, videos or music. The other is a *program* (or application) that enables you to do it.

Data

Data is normally stored on our computers' hard drives or on a device connected to the computer, such as a USB flash memory sick. But it can also be stored in The Cloud via the Internet.

One great advantage of keeping your data on the Cloud is that it is available wherever you happen to be and on any computer to which you have access provided you have your password with you. There is risk however. Organisations that hold your precious data can abandon the scheme or go into liquidation. BT was one of the first companies to offer free cloud storage only later to close it down with little or no warning. There is also some concern about the security of The Cloud. Time will tell.

There is another problem with data. It can become lost, corrupted or hacked. Private information on your computer like passwords can be stolen and misused.

Malwarebytes[108] summarises its views on the subject as follows:

> "Yes, your data is relatively safe in the cloud - likely much more so than on your own hard drive. In addition, files are

---

107 Streaming is listening to music or watching a video immediately instead of downloading a file to your computer to watch later.

108 A supplier of anti-malware.

easy to access and maintain. However, cloud services ultimately put your data in the hands of other people.

If you're not particularly concerned about privacy, then no big whoop. But if you have sensitive data you'd like to keep from prying eyes...probably best to store in a hard drive that remains disconnected from your home computer.

If you're ready to store data on the cloud, we suggest you use a cloud service with multi-factor authentication and encryption. In addition, follow these best practices to help keep your data on the cloud secure:

Use hardcore passwords: Long and randomized passwords should be used for data stored on the cloud. Don't use the same password twice.

Back up files in different cloud accounts: Don't put all your important data in one place.

Practice smart browsing: If you're accessing the cloud on a public computer, remember to log out and never save password info".[109]

Programs (Applications)

In order to be able to create the data the computer needs a **program.** Programs (popularly called apps) are written in code and they customise a computer to do a particular job. For example, a program can have the effect of turning the computer into a super typewriting machine (that enables you to write letters, to email or write a book), into a super calculator (for your accounts), into a photographic studio (for manipulating, displaying and storing photos) or into a music studio and CD/DVD player. The number of different programs that are available is huge and increases daily.

---

109 https://blog.malwarebytes.org/.../should-you-store-your-data-in-the-clou...15 Apr 2016 -

By learning to code, anyone can make an 'app'.

Without having a computer and its various programs, you would need a sophisticated typewriter, a fax machine, an enormous calculator, a photographic laboratory, a CD/DVD player/recorder, a concert venue like the Albert Hall and a collection of books as large as that of the British Library.

The problem with having these programs on your computer is that there are nasty people out there who can use the Internet to interfere with them. Also, they can become corrupted through misuse or due to a faulty computer system. They can also become out of date and may need to be replaced every time computer companies change their Operating Systems. This can be expensive.

Instead of having your own programs on your computer, they can be rented from the Cloud for use when you want them. This means you need a less expensive computer (or use someone else's) and have access to up-to-date and reliable programs (as well as your data) wherever you have access to the Internet.

Big Data

Computer data is like paperwork. It has a habit of growing in size to fill the space available. Much of it is unwanted rubbish but it remains stored. Paper data requires filing cabinets for its storage, which, in turn, attract more paperwork. In a similar way, computer data requires more and bigger disk drives and these attract more stored bytes.

It is said that collectively, stored computer-held digital data is growing exponentially and IBM estimates that by 2020 there will be 43tn [SIC] gigabytes of data; 300 times more information in the world than there was in 2005, all of which is being put to good use. The Internet of Things will be source of much of this data in the future. For example, your energy Smart Meter readings and your web surfing habits are already stored on a computer somewhere.

Some see this global source of data as a goldmine ready to be 'mined'.

Like your computer, the Cloud is also accumulating unwanted data that will never be accessed or deleted. Gartner[110] forecasts that 36% of all consumer digital content will be stored in the cloud by 2016 and that worldwide storage requirement will increase from 329 exabytes in 2011 to 4.1 zettabytes in 2016.

But the term 'Big Data' means more than just storage. A better description would be Big Data Management/Mining/or Processing since it means not only data storage but data processing operations such as analysis, business trends, capture, curation,[111] search, sharing, storage, transfer, visualisation, disease prevention, crime detection, market research, government research, scientific research, medical research and information transfer.

Imagine for example the secure and private storage of the medical details of everyone in Britain. This might include the data transmitted wirelessly from an ECG for example. Programs could then be written to enable any authorised doctor to gain immediate access to this and other to relevant medical data such as DNA profiles and medical conditions. That means that medical specialists throughout the world are then available to give immediate advice and support if required.

An attempt to work towards this end by the NHS has so far fallen short of expectations but one day it will become available.

The Internet of Things

Although not a new concept, there is growing interest in the 'Internet of Things'. IoT means the connection to the Internet of

110 Gartner, Inc. 56 Top Gallant Road, Stamford, CT 06902 USA

111 "Digital curation is the selection, preservation, maintenance, collection and archiving of digital assets" DCC

things, such as domestic appliances and conditions in the home, medical devices, industrial equipment and automobiles, to enable the flow of information to monitor and control them.

Clearly the possibilities for the future are enormous. A surgeon in Australia carrying out robotic keyhole surgery on a patient in the UK using the Internet as a medium is already possible.

The Internet of Me

The IoM is like a personal version of IoT, where it is *us* who are monitored and controlled. Depending on one's point of view this is either the scariest or the most exciting prospect for the future.

When we receive an uninvited email or a pop up that promotes products relating to one that we have just bought or searched for on the web, we are experiencing the early signs of the IoM. It is as though the Internet has stolen our identity and interacts freely with it.

Claims for the future effect of the Internet of Me include Smart system garden monitoring to optimise your specific plants growth, personal sports simulators that track speed, spin, trajectory and behaviour data and then communicate on-line with remote professional trainers, the real-time capture of biometric data for your on-line athletic couching, and fridge content monitoring, automatic shopping and guided cooking to suit your particular taste.

Figure 11 indicates some applications that have been, or will be, influenced by the transistor. Most are candidates for the Internet of Things and the Internet of Me.

Figure 11. The Influence of the Transistor and Future Scope.

<table>
<tr>
<td>

Homes

- Smart thermostats to reduce energy usage with sensors responding to weather conditions, and activities in the home.
- Heating, Ventilating and Air Conditioning (HVAC) systems.
- Smart domestic appliances.
- Security. Response to motion inside the home when vacant and automatically sending a message by email or text when it happens
- Smart lighting.
- Tracking down lost keys, mobile phones and other devices using Bluetooth and other wireless technology.
- Sensors that indicate and track burst water pipes in real-time.
- Systems to feed and water plants based on their actual growing needs and conditions.
- Remote monitoring of washer/dryers using Wi-Fi.

Industry & Retail

- Real-time analyses
- Factory automation
- Robotic management
- Smart Point-of-Sale and dispensing machines
- Point of Sale Manufacturing
- Smart glasses .
- Wearable computers

Aircraft

- Fly by wire
- Communications

</td>
<td>

- Flight data

Infrastructure

- Smart metering
- Smart traffic lights
- Smart parking meters
- Electric vehicle charging stations
- Real-time data analysis.
- Smart grid

Wearables

- Fitness bands
- Smart watches

Automobiles

- Safety
- Vehicle fault anticipation and diagnosis
- In-vehicle Infotainment
- Navigation
- Fleet management
- Electric vehicles
- Driverless Cars

Health and Safety

- Heart monitoring implants
- Ingestible pill sensors
- Biochip transponders on farm animals
- Wearable alarm buttons
- Discrete wireless sensors placed around the home
- A remote 'bins ready to collect' indication to thus reducing the number of pick-ups.
- Sensor systems to remotely read a patient's biometrics (E.G. ECG and heart rate)
- Fire-fighters search and rescue
- Goods tracking in real time

</td>
</tr>
</table>

Your Predictions

I have chanced my arm this far. Next, are few suggestions against which readers might like to add their forecasts and comments. It will then be interesting to turn to this book again in, say, ten years time to see how well you did.

The challenging list reminds us that:

i) The transistor influences almost every aspect of human activity.
ii) Due the influence of the transistor, almost every field of technology is advancing at an exponential rate.
iii) Every available graduate engineer will be in great demand for the long, foreseeable future.

Topics and Comments	**Your Predictions/ Comments.**
Advances in Artificial Intelligence and humanoid robotics.	
Use of Robotics and the effect of robotics on jobs in the UK. Will robots cause hardship by creating unemployment? (By proper planning, robots could generate enough wealth for the country to enable most people to lives devoted to charity work, assisting our neighbours and to enjoying leisure rather than grafting for subsistence).	

3D Printing for the home and for point-of-sale manufacture.	
Computer assisted, more objective and efficient decision-making in government, business and the home.	
Advances in Virtual Reality for simulation, product design and games.	
Further use of Drones.	
More centralised control of devices in the home.	
The use of Genetic Engineering to prevent and cure most or all illnesses.	
Average human life spans? (Increased use of stem cells and gene therapy giving the possibility of indefinite life spans)	
Fault proof smart security methods.	

True 3D T/V without needing special glasses.	
9D television that conveys emotion, movement, pain and temperature and stimulates viewers senses of sight, sound, touch, taste, and smell.	
Printed computers and screens.	
The further use of satellites and greater GPS resolution.	
A single card/token for identity and payments for *all* products and services, loyalty cards etc. using bio-, iris or fingerprint recognition.	
In-car cameras and 'black boxes' as standard.	
A single, shared high street bank location used by all banks in a town; as is now the case with service tills.	

Obsolescence of cash. (Tim Cook, the CEO of Apple, predicts that the children of this generation of students will not know what money is).	
Smart ATMs that not only give access to cash, but also to receive deposits and *instantly* transfer cash and cheques between the various banks. (Australia, Canada and India are all paving the way).	
Improved and widespread use of video conferencing and home working, making travel for work and commuting to and from work uncommon. Fewer office buildings required?	
Future of the cloud.	
Extensions to brains (both human and animals) with the implants of solid-state processors and memory	
Human and animal mind reading and conversing.	
Advanced, high speed and more efficient methods of learning in schools and colleges.	
Artificial eyes and competent artificial inner ears for the blind and deaf.	

Aircraft black boxes supplemented or replaced by real-time transmission to base of *all* flight data for *all* aircraft.	
Further application of simulators.	
The solution to dark matter and an explanation the origin of the universe (or universes). An full understanding of the mysteries of quantum physics. A solution to a theory of everything.	
The possibility of time travel.	
Faultless weather forecasting and control of the climate.	
Air travel at least five times the speed of sound.	
Teleportation. (Even Alice in Wonderland could not have envisaged its possibility yet modern physics is indicating that the phenomenon is possible).	
Progress in the use of Virtual Reality for Education, Science, Medicine and Shopping.	

The High Street of tomorrow.	
Progress on Space Exploration.	
The future use of 3-D printing.	
Driverless cars: Resulting in less accidents, fewer cars, less congestion, lower insurance costs and less street furniture. (Brad Pietras. Vice President, UK Engineering & Technology at Lockheed Martin[112] writes: "It's undeniable that unmanned and autonomous vehicles will play a huge role in the future of our species. This will be both on a practical level as we find ever more inventive ways of removing ourselves from risk and lightening our load, and on a level as autonomous vehicles provide us with an unprecedented ability to explore our own planet and the universe around us. While the future may not look like a scene from a science-fiction movie (complete with talking car), it may well be built with the help of unmanned and autonomous vehicles.")	

112 'E & T' (Engineering & Technology) magazine. Vol.10 Issue 3 April 2015.

Chapter 6. On The Other Hand

All that we have discussed so far has described the huge benefits brought about by the arrival of the transistor. There are those however who are less enthusiastic. They put forward persuasive arguments to show that society is the poorer because of the rapid technological progress over the past decades. Although history reminds us that we cannot stop progress, it would be remiss not to take time to consider the other side of the coin.

Without controlling influences, new products become 'dumbed down' and degraded as the market gives the public what it *wants* without regard to what it *need*s. Television programs, newspapers, web sites and retailed food and drink are just four examples.

There is a fine balance between giving the population its *wants* and its *needs*. In the early days of radio, Lord Reith erred on the side of the latter yet his legacy remains. Critics will argue that we have gone too far the other way. For example, a plethora of radio-broadcasting stations have been made possible by the use of solid-state digital techniques but most of them depend for their existence by pandering to the *wants* of their listeners with wall-to-wall music recordings of little lasting value.

Beneficial inventions become misused. As products become widely used, forces intervene for their own malevolent purposes. For example, the Internet now requires passwords and other forms of protection as safeguards against scams, malware and child access to pornography.

There appears to be a pattern that repeats itself with all new inventions. It goes like this:

> **Stage 1**. The introduction of a new device meets with cynicism and suspicion and condemnation. Its initial cost is outside the reach of most people

The telephone, motor cars, radio, television, and mobile phones have all followed the same pattern. Some 1,000 years ago even the humble table fork was regarded as a device of the devil.

Stage 2. As prices come down and more people use the device there is a period of 'Keeping up with the Jones'. As a result the device becomes generally accepted and its use widespread.

Stage 3. At first, the device is used educationally, beneficially, sparingly and with control.

Stage 5. The device is used by the government, the military and commerce, each grasping its potential for control, power and profit.

Stage 6. The device becomes labelled an anathema and made responsible for all present ills instead of accepting that it is the user, not the device that is responsible.

Here are a few examples that illustrate the views of the critics

The music industry is now dominated by computer-generated music. New sounds like the Hammond organ have been with us for decades but now we have virtual musicians and bands that are almost indistinguishable from the real thing. As an example, orchestral recordings can use fewer human string players augmented by computer-simulated ones. This threatens a falling demand for instrumentalists and fewer young people learning to play string, woodwind and brass instruments.

It is reported that a famous pop guitarist was hired by a recording studio to play just one note on his rare and expensive guitar. The studio then used its computer system to generate a 'single' with the name of the famous guitar player prominently displayed on the label. Apparently, it became a best seller.

Manufacturers contrive to keep their customers glued to their computers, televisions and mobile phones. Yet, health advisors, urge us to be physically active. For example James Titcomb, describing the launch of Apple's television, says,

> "Thc argument is that we are moving ever further from the living room. But Apple hopes its new box's multimedia properties will keep us glued to our sofas".[113]

There are signs that the fast pace of modern living, brought about by the transistor and its derivatives, such as the computer and the smart phone, is increasing stress and unhappiness. According to the Daily Telegraph,[114]

> "... in an era of mobile communications .... millions of us feel we have less direct contact with other people."

Although the mobile phone system gives many benefits, it has led to the erection of unsightly relay towers with their possible health hazards.

It is anticipated that there will be a huge rise in deafness in twenty years time. This is attributed to subjection to loud headphones and high power, solid-state disco amplifiers when young. No doubt solid-state will be blamed for this instead of the misuse of the technology.

The low cost and small size of programmable silicon chips has led to over-complicated products with features that are rarely or never used. The usefulness of some of these facilities is questionable. Manufacturers may include them in order to upstage their competitors. For example, cameras are sold with resolutions of over 20 mega pixels yet, 3 mega pixels is more than adequate for a 6"x 4" print. Camcorders have storage for video recording durations equal to twice the number of

---

113 The Daily Telegraph 'Business'. Friday 11 September 2015
114 17th April 2015

gigabits divided by the mbps[115] This can amount to several hundred hours, yet very few users need to keep that quantity on their machines.

Washing machines and dish washers have several settings that users often neither need nor bother to understand. It reminds me of a Blue Streak brainstorming session to seek ideas for monitoring the vertical orientation of the rocket on the launch pad. Ideas ran wild as increasingly complicated circuits with escalating numbers of transistors were proposed. Then a small voice at the back of the room suggested a weight suspended on the end of a piece of string. His solution was accepted and proved quite satisfactory.

The Internet, the derivative of the US government's Arpanet, began with a relatively few, but immensely useful, web domains mostly devoted to educational material. Soon, commercial interests dominated the web and, because money exchange was involved, it attracted all forms of malware and fraudulent behaviour. Rogues, for whatever twisted reasons, began to devise viruses to disrupt the web and impair the operation of computers. The stock market, government, utilities and commercial computer systems have become vulnerable to catastrophic damage because of 'hacking', as experienced by Talk Talk in 2015. There are however legitimate hackers who are employed to help counter the terrorists by playing them at their own game.

The advent of Quantum computing will bring unheard of benefits to science, health and daily life. However there are potential dangers. For example, we use our credit cards safely in the knowledge that the encryption uses public-key systems with prime numbers so large that they are impregnable by present-day computers. It is said that calculating the private key from the public key would take longer than the expected

115 Megabits per second; the picture quality.

lifespan of the solar system. But quantum computers could do the job in a realistic period of time. Will we ever solve the problem of data security?

Advances in Artificial Intelligence and Robotics will have a profound effect on work. Professor Mosche Vardi[116] considers that they could cause unemployment rates within the next 30 years of 50 percent. But is there a brighter side to Professor Vardi's prediction. If sufficient productivity is achieved, early retirement, even unemployment, could give a life of freedom financed by the country's Gross National Product. We could then devote our lives to voluntary work, helping others and taking part in healthy leisure activities.

The transistor has given us products that are complex and cheap causing us to throw away and replace, rather than to repair them. This has resulted in overuse of natural resources, loss of DIY repair skills and the growth of the recycling industry.

Opposing voices are usually those that do not understand the technology and are therefore suspicious of its effects. A legendary example is that of a person who worried that leaving a light socket without a bulb would allow the electricity to leak out. (Understandable if you have been used to gas lighting). These voices condemn progress by what they see around them rather than seeking the true cause of their concern. The frustrations and dangers of the Internet for example are because others misuse it. We would not have arsonist had fire not been discovered. We would not have great guns and tanks if there had been no iron age, but nor would we have cures for polio, AIDS and other dreadful diseases if we had stayed with bloodletting and magic potions.

116 Professor of Computer Science at Rice University, USA

We cannot stop progress but we can control it. Caring members of society can avoid trailing behind progress and take the lead to help humankind adapt to the changes and restrain bad behaviour.
Above all, scientists and engineers must strive to find ways to counter the misuse of their endeavours and to make their work intelligible to everyone.

Epilogue

We have come a long way from the heroic struggles described in Chapter 1, from the imagination, perseverance and genius, of men and women working alone and hampered by primitive materials and resources. Many of them are unrecognised yet they laid foundations for others on which to build.

Now, after little more than 100 years of unprecedented progress, we live in a world that even sc-fi writers find it hard to outdo - and we take it for granted.

The Hydron Collider, the A380 Jumbo Jet, supercomputers, space travel, space telescopes, our knowledge of quantum physics and the universe, and awesome advances in medicine compare favourably with fantasy. They are all advances that would not have been possible without the invention of the transistor, only 70 years old in the year 2017.

It is a characteristic of our time that the scientists and engineers who work behind the scenes to bring about these astonishing advances are overshadowed by the front line operatives who make use of them. For example, airline pilots, train drivers, surgeons, users of domestic appliances and those in commerce (all no less competent people). It is hoped that this book has helped to restore that balance a little.

Arguably, no other 'backroom' like that at Bell labs in 1947 has produced a device that has transformed our world so dramatically in such a short period of time. Even John Bardeen, William Shockley and Walter Brattain would be amazed at the changes that their transistor has brought about. They proved that Charles H Duell's[117] prediction; "Everything that can be invented has been invented" is as wrong today as it was in 1899.

---

117 Commissioner of the United States Office of Patents.

The work of Ernest Duchesne, a French physician, illustrates another factor that is relevant to our story. In 1897 Duchesne, at École du Service de Santé Militaire in Lyons, studied the interaction between Escherichia coli and Penicillium glaucum. He independently discovered the healing properties of *P.* glaucum, even curing infected guinea pigs from typhoid. However, the Institut Pasteur, his employer, ignored Duchesne's dissertation about the discovery so we had to wait decades for penicillin. This episode shows that many of us may have ideas and fertile imaginations but success comes about by actually doing something about them,[118] and for the benefit of humankind.

Let us imagine a nation that, unlike the other nations in the world, had never heard of the transistor or it progenies. By the twenty first century, that nation would be just a forgotten backwater or would no longer exist. Such is the massive scale of influence that a tiny semiconductor crystal has had on the world.

The transistor has helped us to make sophisticated weapons that can kill more people but it has also helped us to discover the human genome that enables us to cure people and save lives. It is for us to learn where our priorities lie. Let us learn to be grateful to the Bell Labs team's invention in1947, not regret it.

Hopefully solid-state technology will further develop the means to settle our differences in a way that does not kill people or create destruction - a logical system of independent, fail-safe diplomacy.

Whatever opinion we have of him, I think Albert Speer's final speech in his defence at the Nuremberg Trials is an appropriate way to end:

> "The more technological the world becomes, the greater the danger ...... There is nothing to stop unleashing technology and science from completing its work of destroying man which it has

---

118 Like John Logie Baird, Thomas Edison and Alexander Fleming for example.

so terribly begun in this war .... The nightmare shared by many people, that some day the nations of the world may be dominated by technology - that nightmare was very nearly made a reality under Hitler's authoritarian system. Every country in the world today faces the danger of being terrorised by technology; but in a modern dictatorship, this seems to me to be unavoidable. Therefore, the more technological the world becomes, the more essential will be demand for individual freedom and the self-awareness of the individual human being as a counterpoise to technology .... Consequently this trial must contribute to laying down the ground rules for life in human society. What does my own fate signify, after all that has happened and in comparison with so important a goal?"

An Afterword

Writing this book brought back memories of many happy and satisfying times in my career as an electronic engineer.

Hopefully, 'The Transistor' will encourage others to take up the challenges offered by a tantalising future.

I much appreciate your time in reading the book.

Appendices

Appendix 1 The Radio Wave Carrier

It is not obvious how the waves of music, speech or data and the electromagnetic waves of a radio transmission are related. It is because of a process called modulation.

Modulation is the method of impressing the music, speech or data signal (sound or data) waves on to an electromagnetic radio (carrier) wave.

There are a number of ways to achieve this modulation. The two that were most widely known in 1950 were Amplitude Modulation (AM), where the carrier wave's *amplitude* is varied by the sound or data wave and Frequency Modulation (FM), where the carrier's *frequency* is varied by the sound or data wave. These are illustrated below.

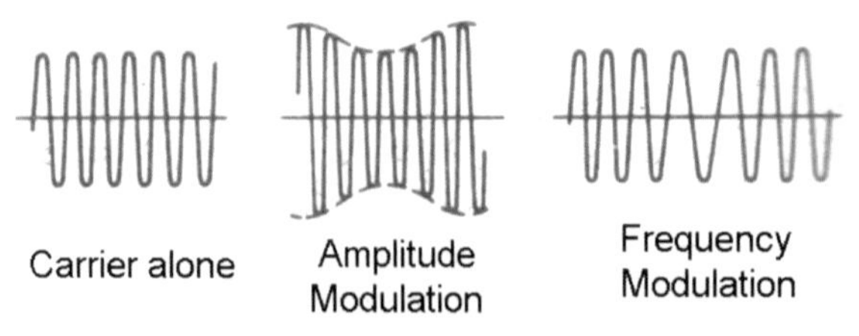

An electromagnetic wave carrier modulated by a sign wave.

Appendix 2. The Transistor

Without being armed with the necessary mathematics, any verbal description of the transistor is liable to be inadequate and open to ambiguity; anyone with an enquiring mind will always be left with further questions. However, let us try by making one. (In theory of course, otherwise you will need the skill, the patience and several million pounds to spare for the tooling and materials).

First, we will make a *diode.* To do this, we need a PN junction as follows:

a) Take a piece of a semiconductor. We will use silicon because, among other reasons, it is a chemical element commonly found in sand and is therefore available in almost unlimited quantity. Now refine it. Then take the refined piece and grow it into a very, very pure crystal.

b) Add a controlled amount of antimony as an impurity to 'dope' each silicon atom in the piece of pure silicon. You have now made an N type semiconductor because it has *free electrons* to give away. At this stage, this newly doped N-type semiconductor crystal does very little because it is electrically neutral.

c) Now, we need to form a *'PN Junction'* in the N-type crystal. There are a number of ways to do this but we will use the oldest method, because it is the easiest to understand, so we will form an 'alloy junction'. First, we put a small quantity of aluminium on to the surface of the N type semiconductor crystal and, using a furnace with an inert atmosphere, we heat it up to the temperature of the Eutectic Point. (Its lowest possible melting point). A thin film of melted aluminium then forms on the surface. As it cools, it all solidifies into a single crystal, partly P type semiconductor and partly N type semiconductor material.

We used aluminum because it has an excess of *holes*. Holes are the absences of free electrons.

It is important to note that this is not the same process as joining two metals together. We are fusing them onto one crystal. However, a thin barrier layer exists at the junction between the N and P-type semiconductors. (See Figure 12).

Figure 12. The Construction of the Diode.

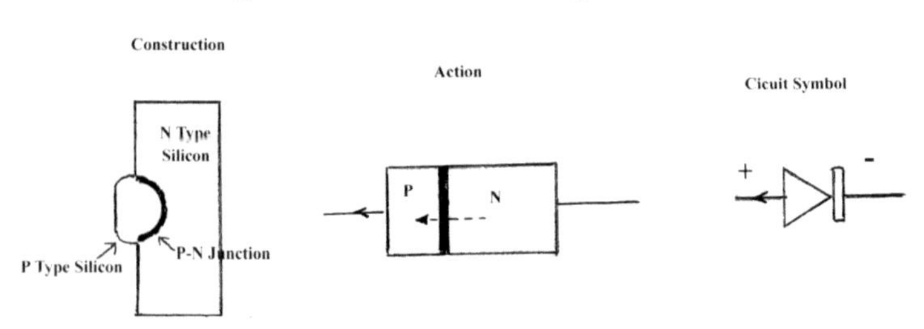

d) When the N-type semiconductor and P-type semiconductor materials are first fused together free electrons and holes diffuse each way across the junction and jostle for partners. I bit like very fast Speed Dating. Eventually an electrically neutral situation settles leaving a "potential barrier" zone around the area of the P-N junction. This is called the Depletion Layer.

The Depletion Layer has a very high resistance to electric current but it also gives the crystal a special characteristic. If a voltage is applied to the crystal, in one direction, it will remain a high resistance and no current will flow. But if the voltage is applied to the crystal the other way round it will present a low resistance and current will flow. Put simply, if a positive voltage is applied to the P region (the anode), electrons are allowed to cross the P-N junction from the N region. Then they then have nowhere else to go than through and out of the P region.

e) Finally, we need to attach leads to the P and N regions of the crystal and enclose it in a hermetically sealed case with two protruding leads labelled 'anode' and 'cathode'. Eureka! We have made a diode! (Figure 12)

Now we will make a bipolar junction *transistor*.

For this we need to construct a crystal that comprises two diodes back-to-back, each formed by the method we used for the diode.

The resulting N-P-N crystal resembles a sandwich. Two 'slices of bread' of unequal size are of N material called the *collector* and the *emitter*. The sandwich is filled with a P-Type layer called the *base*. Where each of the slices meet the filling there will be a junction – a depletion layer - similar to the one we saw with the diode. (See Figure 13).

Figure 13. Construction of the Bipolar Junction Transistor.

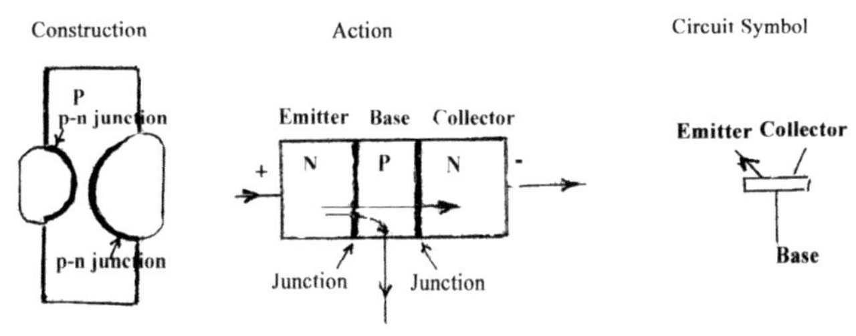

If a voltage is applied to the base, that is positive relative to the emitter, electrons cross the P-N junction from the emitter, just as we saw with the diode. However, if we now also apply a larger positive voltage between the collector and the emitter most of the electrons that cross the P region are pulled across the collector/base junction. This is encouraged because the P region is very narrow and the collector/base junction is large. We might imagine the collector/base junction as behaving like a waterfall.

Because of this 'theft' of electrons by the collector, very few of them are available to flow out of the base. Hence, a small current through the base results in a large current though the collector.

All we have to do now is to attach leads to the three regions of the crystal and enclose it in a hermetically sealed case with the three protruding leads labeled emitter, base and collector.

We now have a transistor!

In practice, one does not need to know any of this science about electrons and holes unless their purpose is to design semiconductor chips. All the circuit designer needs to know is that a transistor works like an amplifier or a switch. That is, it uses a small emitter to base current to control a large emitter to collector current. This is like varying the lights in a room or turning them on and off using a simple switch control.

Various types of transistor, and passive components like resistors, capacitors and inductors, can then be wired together to build circuits for particular applications.

Before leaving this section, it may be useful if I risk confusing the reader with a water analogy of how a transistor works..

Referring to figure 14a imagine a tank **D** with a sloping bottom and two sluice gates. The smaller one on the right **G** is analogous to the emitter of the transistor **B** and the larger on the left **F** to the collector **A**. The drain hole **C** at the bottom is equivalent to the base.

Figure 14a.

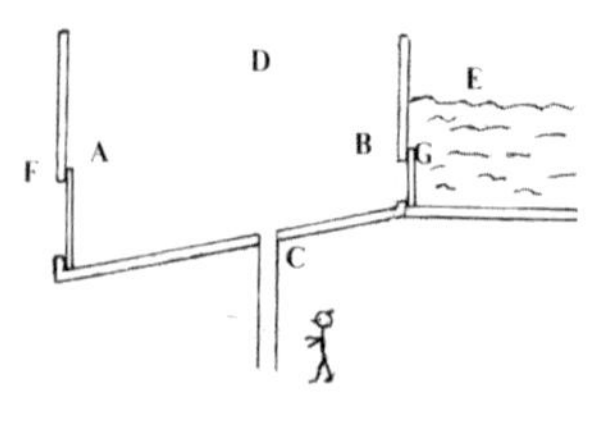

On the emitter side of the tank is a source of water **E** representing electrons. An operator (the input controller) is standing near the base waiting to open the emitter sluice gate. Both sluice gates are shut and no water (electrons) can get through. That is, our transistor is not connected.

Figure 14b

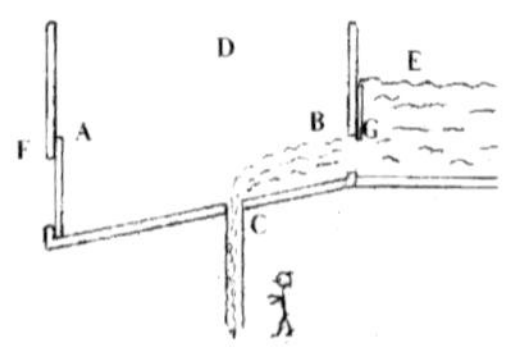

The operator at the base has opened the emitter sluice gate **G** and water rushes in. It has nowhere to go except down the drain. (Like current through the base that has a positive voltage applied). The water can only go in one direction so it acts like a diode.

Figure 14c.

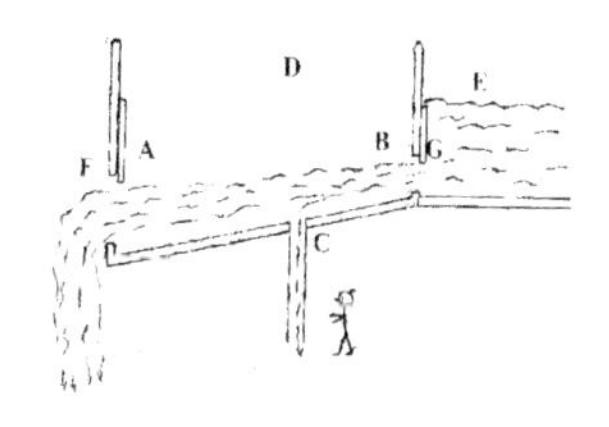

Now the collector sluice gate **A** is opened (Like a positive voltage being applied to the collector) and the operator has opened the emitter sluice gate. Water rushes in and, due to its momentum and the larger opening of the collector, most of it passes (drifts) through and very little travels down the drain (the base). So, a small action by the controller has caused a large action from the emitter to the collector. The controller could have opened and closed the sluice gate completely to represent a switch or continuously, to represent an amplifier.

Appendix 3. The Transistor's Progenies

The transistor has spawned several other solid-state electronic components, each with specialised characteristics. Here are just some that are available to the electronic engineer.

Bipolar transistor NPN and PNP
MOSFET Transistor
JFET Transistor
HEMT Transistor
IGBT Transistor
UJT Transistor
Darlington Transistor
Silicon Controlled Rectifier (SIC)
Thyristor
TRIAC
SIDAC
DIAC
Light Emitting Diode (LED)
Photo Sensitive Diode (Opto)
Constant Current Diode
PIN Diode
Schottky Diode
TVS Diode
Varactor Diode
Bridge Rectifier
Zener Diode

Appendix 4. A Further Word About Programming

Programming has been used at least as far back as 1801 when Jaquard coded his programs for his Loom by punching a series of holes in cards. Also, we have all heard a fairground organ with its strips of card perforated with holes that are the codes for the musical notes and registration controls for the organ. It was known as 'book music'.

The beauty of these machines is that they could be made to operate differently by the choice of a card containing the coding for a particular cloth pattern or a piece of music. The mechanics of the machines did not have to be changed. It was an unfamiliar concept that people found hard to grasp because, hitherto, different machines were built to do different jobs.

It was not until the transistor arrived that programming infiltrated almost every aspect of science and technology. Apart from the transistor itself, programming the *general-purpose* digital computer has probably had more influence on progress than anything else.

With a personal computer or laptop to hand, let us consider a few things where programming changes the personality of this, just one, machine to do the jobs of many.

We can throw away our typewriters and specialised typewriter/word processors because programs like Word, WordPerfect and Open Office Writer on a computer do the job much better. A Spellchecker and a Thesaurus are included so we can also ditch our dictionary and thesaurus books.

We can also throw away our desktop calculators and quill pens because computer programs like Excel, Apple Numbers or Open Office Calc do a much better job.

We can throw away our FAX machine because our computer system can scan and email documents using programs like Outlook, Thunderbird or Eudora.

We can stay away from typing classes because there are several programs available to teach us, that are more convenient. We can even dictate speech to the computer.

There are many powerful programs, like Photoshop, that make it easy to modify, enhance or restore our photos.

We can throw away our CD and DVD players because our computer can play them just as well.

There are also many less known programs. For example:

We no longer need manuscript paper on which to write down our music. Instead, we can use a computer program to notate, print, play back, transpose and generate the parts for the instruments. For example see:
https://www.finalemusic.com/
Or
http://www.sibelius.com/home/index_flash.html
Then we can realistically imitate a 17-piece Big Band, a 40-piece orchestra to play our masterpiece on our computer.
See/hear: *http://www.garritan.com/*

Our computer can accurately imitate the fine cathedral organs of the world to play in our living room.
See/hear: *https://www.hauptwerk.com/instruments/*

We can discard our drawing board and drawing tools because we can do the job much more quickly using programs like AutoCAD and OrCAD. We do not need a blueprint machine because the computer has a program that sends our work to a printer.

A program on our laptop could control several looms. It could hold many thousands of patterns and any one of them can be selected and produced by the click of a mouse.

We no longer need a map book to plan our journeys. A program on our laptop (or mobile phone) will work out and display the journey for us, give the travel times and distances and tell us what to do

We can install a variety of programs on our laptop that can give realistic simulations that teach us to fly a plane, drive a car, drive a train, run a farm and even fight a war without incurring real casualties.

There are several languages for writing programs. Some, like Basic, Visual Basic, and HTML can be learned from books without much difficulty.

Bibliography

'A History of Early Television'. Volume 1. 2004. Stephen Herbert. ISBN 0-415-32666-4
Alan Turing. On Computable Numbers, with an Application to the *Entscheidungsproblem*" (submitted on 28 May 1936 and delivered 12 November).
'Atomic. The First War of Physics'. Icon Books Ltd. 2009. ISBN 978-184831-044-5
Chief Superintendent of the Telecommunications Research Establishment (TRE) 1938-1945. Camb. University Press 1948
Commissioner of the United States Office of Patents
Computer Weekly.com
Computing With Quantum Cats. John Gribbin
Daily Telegraph 22nd Warner, 6th January 2016.
'Flight 100 Years of Aviation' RG Grant. Dorling kindersley Limited 2010.
'From the Ground Up'. F.J.Adkins. Airlife Publishing Ltd.1983 ISBN 0 906393 21 3
Gartner, Inc. 56 Top Gallant Road, Stamford, CT 06902 USA' E & T' (Engineering & Technology) magazine. Vol.10 Issue 3 April 2015
http://www.computermuseum.org.uk/fixed pages/Elliott 903.html
http://www.graphenea.com/pages/graphene-properties#.Vczqz mumeB
http://www.theguardian.com/sustainable-business/big-data-impact-sustainable-business
https://en.wikipedia.org/wiki/Raspberry Pi
https://en.wikipedia.org/wiki/Transistor count#cite note-33
Irving, David (1964). *The Mare's Nest*. London: William Kimber and Co.
'Jim Austin Computer Collection'. One World Publications 2010. ISBN 978-1-85168-779-4. Page 34
Journal of the Brit.I.R.E. January 1961
'Mathematical Work of Charles Babbage'. John Michael Dubbey. (Cambridge, MA: Cambridge University Press, 1978)
'Modulation'. 2nd Edition. F.R.Conner
Peter Hammer. MIX 1999
Simon Lavington (Editor.), 'Alan Turing and His Contemporaries.
Skrabec, Quentin R., Jr. (2012). The 100 Most Significant Events in American Business: An Encyclopedia . ABC-CLIO. p. 197. ISBN 0313398631.
'The Proud Tower' by Barbara W. Tuchman. Hamish Hamilton 1996.
'The Secret Life of Bletchley Park'. Aurum Press Ltd. ISBN 9781781315866
The Radio and Electronic Engineer 1967.
The Radio and Electronic Engineer. Vol 38. No.6 December 1969.
'The Story of the Laser'. John M. Carrol.
'The Shorter Oxford English Dictionary'
'The Giant Book of Bathroom Dumbology'. Geoff Tibballs. Constable, London. 2014
'Vulcan Test Pilot'. Tony Blackman. Bounty Books 2015
'Waves, particles, and atoms'. Nuffield Advaonced Science. Penquin Books Ltd. 1971
Wikipedia. 'History of multitrack recording' 8 August 2014

Index

D

E

O

P

Q

R

S

T

U

V

W

X

Z